Palgrave Studies in Climate Resilient Societies

Series Editor
Robert C. Brears
Avonhead, Canterbury, New Zealand

The Palgrave Studies in Climate Resilient Societies series provides readers with an understanding of what the terms **resilience and climate resilient** societies mean; the best practices and lessons learnt from various governments, in both non-OECD and OECD countries, implementing climate resilience policies (in other words what is 'desirable' or 'undesirable' when building climate resilient societies); an understanding of what a resilient society potentially looks like; knowledge of when resilience building requires slow transitions or rapid transformations; and knowledge on how governments can create coherent, forward-looking and flexible policy innovations to build climate resilient societies that: support the conservation of ecosystems; promote the sustainable use of natural resources; encourage sustainable practices and management systems; develop resilient and inclusive communities; ensure economic growth; and protect health and livelihoods from climatic extremes.

More information about this series at
http://www.palgrave.com/gp/series/15853

Zaheer Allam · David Jones · Meelan Thondoo

Cities and Climate Change

Climate Policy, Economic Resilience and Urban Sustainability

palgrave
macmillan

Zaheer Allam
Live + Smart Research Lab
School of Architecture and Built Environment
Deakin University
Geelong, VIC, Australia

David Jones
School of Architecture and Built Environment
Deakin University
Geelong, VIC, Australia

Meelan Thondoo
Centre for Research in Environmental Epidemiology
Barcelona Institute for Global Health
Barcelona, Spain

ISSN 2523-8124 ISSN 2523-8132 (electronic)
Palgrave Studies in Climate Resilient Societies
ISBN 978-3-030-40726-1 ISBN 978-3-030-40727-8 (eBook)
https://doi.org/10.1007/978-3-030-40727-8

This Palgrave Macmillan imprint is published by the registered company Springer Nature Switzerland AG
The registered company address is: Gewerbestrasse 11, 6330 Cham, Switzerland

Foreword

Humankind has a rich history, and as the story of man and its relationship with their world unfolds, we are taken into episodes of conquests, pain, and love. But just like any story has a takeaway, ours will most definitely be that of survival. And I truly believe this—and maybe this is my optimistic nature, that we are a triumphant society, and just like we did numerous times through the episodes of war, of famine, or of crisis, we persevered, and we survived! So today, when we talk about climate change as the single most critical event of our lifetime, we need to take a few steps back and look at our history from a critical lens, we need to acknowledge the dire consequences—the science behind it, and we need to also take into account the incredible resilience of humankind. While I will personally not be here to witness this act of resilience, I will ask of the next generations to be hopeful in the future and to trust the ability of humankind for survival. But of course, this does not mean that while knowing that we are living in an era of crisis, that we can carry on business as usual. This act of survival starts now!

I have had a long and rich career in policy making at both national and international level, and have had the opportunity to travel to countless conferences—which by now must be in hundreds, and I have seen the slow but steady shift of topics addressed to include more sustainably inclined agendas. Of course, I often say that sustainability in itself is not new or can be interpreted as being novel, as in reality it has been on the agenda for over 30 years, since the establishment of the Intergovernmental Panel for Climate Change (IPCC). And I often, even though my diplomatic training, find it frustrating that sometimes we end up reiterating the same arguments in conferences or meetings globally, and I at times call for more actions rather than more resolutions. But to be honest, the process of change is an important of, as disruption calls for conflicts and inequalities. So, as an experienced Politian, and someone with an experience in the processes of the United Nations, speaking about resolutions and reiterating those conversations are important too for the sake of global prosperity. Ultimately, it is those processes that trickle down to action at local level, but acknowledging that this process takes time, we can also pre-empt and work on our own localised policies. This is where we need good politicians that understand the crisis of our time, the geopolitical context, and local economics. The last point is key, as I truly believe that is it through economics that we can drive change; even environmental.

So, what are the economics of survival? And how can economics be applied in the field of environment while keeping the global agendas of prosperity that those lengthy resolutions support? This is where I believe our innovative spirit is needed, and where we need to break away from our resistance to change. As if we want to address climate change, and if we look at the economics, there is a lot of money to be made while saving the planet. It is not only a philanthropic endeavour. We see how the sales of renewable energy is trumping that from coal, the sales of electric cars dethroning petrol ones, amongst others. So why not join this shift where we can not only make money, but also help in the redistribution of wealth through the disruption of global monopolies. What is interesting here is that we are seeing action on climate change also equating to action towards bridging the gap of social disparities; which is the key in sustaining healthy communities. Having witnessed this shift of innovative disruption from our traditional linear thinking in a matter of two decades is what, as a politician and diplomat, makes me hopeful for the future.

We need more of this hope, we need to encourage more innovative solutions, and we need more people working on disrupting their respective industries.

The book of Zaheer, Meelan and David is one of hope, and one that underlines the human spirit of survival. They dwell into our world of climate challenges and suggest climate policies that have for ability to provide solutions applicable at both local and global scales through a trans and cross disciplinary perspective. And they are not alone, collectively this wealth of information will help shape a better world, better societies, and healthier communities. This book will help in building a more humane world and help reiterate the nature of humankind: our optimist nature of survival, even when faced with the dire challenges of global warming.

So, let's join them and call for action. In doing this, we need to also recognise the need for intelligent action that plugs in our global economic models. Winning the climate change debacle is possible, but we need to proceed with prudence. We need to proceed strategically. That's the advice I would give.

Pereybere, Mauritius Armoogum Parsuramen

Armoogum Parsuramen Minister of Education, Arts, and Culture and Minister of Education and Science of Mauritius both between 1983 and 1995 and Director and Representative of UNESCO between 1998 and 2011, Armoogum Parsuramen is currently the founder-President of University of 3rd Age and Global Rainbow Foundation (GRF). For his services to the Mauritian Society, he was elevated to the rank of Grand Officer of the Order of the Star and Key of the Indian Ocean (GOSK) by the President of Mauritius.

Contents

About the Authors

Zaheer Allam is a holder of a Ph.D. from Curtin University (Australia), an M.A. in Political Economy from University of Sydney (Australia), an M.B.A. from Anglia Ruskin University (UK) and a Bachelor of Applied Science in Architectural Science from Curtin University (Australia). Based in Mauritius, he works as an Urban Strategist for The Port Louis Development Initiative (PLDI), the Global Creative Leadership Initiative and consults on a number of projects on the thematic of Smart Cities across the African Continent and on strategies dwelling in the increasing role of technology in Culture and the Society. Zaheer is also the African Representative of the International Society of Biourbanism (ISB), member of the Advisory Circle of the International Federation of Landscape Architects (IFLA), and a member of a number of other international bodies. For his contributions to society, he was made recipient of a number of awards and was elevated, by the President of Mauritius, to the rank of Officer of the Order of the Star and Key of the Indian Ocean (OSK); the highest distinct order of Merit in Mauritius. He is the author of *Cities and the Digital Revolution: Aligning Technology and Humanity* (2020) and *Theology and Urban Sustainability* (2019).

David Jones is the Foundation Professor of Landscape Architecture and Planning at Deakin University since 2011. He has degrees in planning, landscape architecture and heritage studies, including a Ph.D. from the

University of Pennsylvania. His teaching and research spans across Urban Planning, Landscape Architecture, Indigenous Knowledge Systems, Regenerative Systems and Biophilia. His portfolio includes work on the Forest Gallery at Museum Victoria (1995–1996), the Victoria Square—Tarntanyangga Regeneration Project (2017), the *Adelaide Park Lands and Squares Cultural Landscape Assessment Study* (2007), and on design scenarios for Gunditjmara lands associated with the Budj Bim National Heritage Landscape and their World Heritage Listed property. He is the co-author of *Geelong's Changing Landscape: Ecology, Development and Conservation* (2019), *Re-casting Terra Nullius Blindness* (2017), *Creating Healthy Places: Railway Stations, Biophilic Design and the Melbourne Metro Rail Project* (2017), *Aboriginal Reconnections* (2013), and has co-contributed significant chapters to the *Routledge Handbook to Landscape and Food* (2018) and *The Handbook of Contemporary Indigenous Architecture* (2018).

Meelan Thondoo is currently completing a joint Ph.D. in Medicine and Anthropology, with a focus on developing a sustainable future for vulnerable populations, in affiliation with the European Commission and leading universities. She is developing the first framework for Health Impact Assessment (HIA) for Urban and Transport Planning in fast developing cities and resource-constrained countries of Africa. Meelan holds an M.Sc. in Medical Anthropology and a MPH in Health Economics from University College London and the London School of Hygiene and Tropical Medicine respectively. Her commitment to promoting health in local and global communities, across three continents is translated through affiliation with and board memberships on international networks and multilateral bodies such as the World Health Organization, DFID UK and Bill and Melinda Gates Foundation. She has received academic and pro-bono awards from the University of Salamanca, the World Bank Institute, the UN Global Humanitarian Forum and she is a fellow of the Royal Anthropological Institute.

List of Figures

CHAPTER 1

Urban Resilience and Climate Change

Abstract As the world witnesses a significant demographic boom, the impacts of climate change are getting more pronounced and affecting the livelihoods of people as well as urban ecosystems from climatic, health, environmental and economic standpoints. The role of cities is apparent in this process. Cities are now shown to contribute to urban, national and regional economics driving different dimensions of world development. Today, cities are designed to respond to the needs of contemporary urban economics, and there is an increasing literature calling for new concepts aligned with more sustainable outcomes. However, the implementation of new concepts will mean restructuring a number of vital urban infrastructures, which comes at a cost. Even though the need for those is now acknowledged, there is no consensus as to what models are readily available to support this change.

Keywords Urban resilience · Climate change · Sustainability · Cities · Economics · Liveability and health

Z. Allam et al., *Cities and Climate Change*,
Palgrave Studies in Climate Resilient Societies,
https://doi.org/10.1007/978-3-030-40727-8_1

Introduction

From the nineteenth century, the world's population has maintained a steady exponential growth. Today, as humans we comprise 7.6 million residents on this planet, and by 2050 it is projected that this will continue to rise to an estimated 9.9 billion people (United Nations, 2017). What is surprising about this increase in population is that a majority of the world's population is opting to live in urban centres; thus, fuelling the global phenomenon of urbanization. From records, of the 7.6 billion people, 54% are currently living in urbanized areas and with every day, more are migrating or being born in cities. Thus, it is projected that by 2050 that 68% of the world population will be living in cities (Population Reference Bureau, 2018). The result of this trend is the expansion and unfettered sprawl of cities, leading to the consumption of green and agriculturally productive spaces at the peripheries of urban areas that are engulfed in seas of concrete, tin and plastic. Unfortunately, such extensions of urban boundaries are compromising land reserves that seek to protect resources and assets like forests, parks, waterfronts, coastlines and green spaces amongst others that provide the 'green lungs' for these cities and contributing oxygen to cleanse cities; improving human and animal well-being (Allam, 2012, 2017; Allam, Dhunny, Siew, & Jones, 2018).

Another intriguing consequence of this increasing population and high rate of urbanization is the parallel demand for a myriad of services such as energy, health services, consumer goods, infrastructural development, housing to service urbanization and human needs (Corey, Wilson, & Fan, 2015; Lang, 2018; Zhu, 2017). Indeed, the world is now characterized by a highly consumerist population that is exerting unprecedented pressure upon available resources. This is perhaps well represented in the desolating images of the overexploitation of forests (Figs. 1.1 and 1.2).

The most unfortunate outcome of these patterns is that our lifestyle and global trends are accelerating the decline of environmental sustainability and impacting upon the liveability of urban areas, more so with the advent and consequences of climate change. And, without any doubt, climate change is now documented as one of the biggest threats of this century to planet Earth and its residents, with many impacts being felt in most cities globally, irrespective of their nation's development status (Mgbemene, Nnaji, & Nwozor, 2016; OECD, 2014; Rezai, Taylor, & Foley, 2018). These consequences include unprecedented flooding (Arnone, Pumo, Francipane, Loggia, & Noto, 2018; Miller & Hutchins,

Fig. 1.1 Deforestation (Image by Dave Herring)

Fig. 1.2 Dense forest in Sao Paolo (Image by Sergio Souza)

2017), extreme increases and drops in temperatures, high precipitation levels and significant erratic climatic incidents, emergence and increase of new traits of diseases (Ogden, 2018) and prolonged droughts and dry spells amongst others (Ghebrezgabher, Yang, & Yang, 2016). Surprisingly, these events are also being felt and affecting remote areas that do not boast high urbanization rates or engage in high pollution-intensive activities which were thought to be safe from certain forms of weather disasters. A case in point is the case of cyclones Kenneth and Idai that hit countries such as Mozambique, Malawi and Zimbabwe on 4–21 March 2019 affecting numerous people and leaving behind trails of massive destruction of in excess of $2.2 billion (2019 USD) (Onishi & Moyo, 2019). Such cyclone incidents had historically not been reported in this region.

In cities, most of which are now densely populated (Figs. 1.3 and 1.4), the consequences of climate change on vital urban infrastructures have also been reported to cause immeasurable havoc and chaos (Allam, 2019a, 2019b, 2020e, 2020f, 2020g; Allam & Jones, 2018, 2019).

Effects include claiming lives of millions and negatively affecting their livelihoods and impacting upon entire urban economies (UN-Habitat, 2015; World Bank, 2010). For instance, such destructive events like hurricanes Katrina (2005), Harvey (2017), Jebi (2018) and Idai (2019), and the numerous wildfires that are being experienced in different countries

Fig. 1.3 Densely populated city of Sofia in Bulgaria (Image by Georgi Kalaydzhiev)

Fig. 1.4 City of Kuala Lumpur, Malaysia (Image by Ishan @seefromthesky)

have had severe impacts on urban infrastructures (BBC News, 2018). Roads, bridges, power plants and distribution lines, water supply systems and airports have been reported destroyed, eroded, flooded and laid waste leaving substantial parts irrevocably damaged. For instance, in 2017, storms, hurricanes and cyclones alone were responsible for an accumulated destruction amounting to over $300 billion in the Atlantic Basin, and the displacement of millions of people with over 3500 reported deaths (NOAA, 2018, 2019).

The reality of such destruction and loss is that climate change is now perceived to be a global threat sparking concerted policy and action efforts from global major players like governments, international organizations, institutions and diverse stakeholders to propose urgent mitigation strategies (Condon, Cavens, & Miller, 2009). In particular, in cities, there are concerted citizenry calls for revising urban planning models and policies to better accommodate practices and projects that promote increased resilience against climate change (Mi et al., 2019). However, the foundational operating processes of urban life are being questioned as cities are major contributors of climate change. To a larger extent, one has to agree with the critics since the current urban planning models are not ecologically inclined and not adapted to the current needs (Apreda, D'Ambrosio, & Di Martino, 2019), in their capacity in accommodation and successfully guiding responses to current issues including increasing urban population and consumption behaviours to list a few. Our consumption, especially of energy from non-renewable sources, has been identified as possessing a high propensity to increase emissions thereby negatively accentuating climate change impacts. In addition, demand for resources like foodstuffs, water and building materials exacerbated by increasing urban populations also plays a key role in climate change because of their extraction and fabrication needs prompt unhealthy practices including wholesale vegetation clearance and firing of more land to facilitate agricultural productivity.

Such activities lead to overexploitation of resources like water to irrigate agricultural lands, excessive soil erosion and transformations of biodiversities into monocultures to reputedly produce enough food supply to satisfy our insatiable urban populations (Wiebe, Robinson, & Cattaneo, 2019). The constant call to action on climate change has resulted in the formulation of numerous policies at international levels targeting governments and other stakeholders who are directly involved in not impeded these activities. The weight of these policies is for international and urban players to emphasize sustainable practices in various spheres of their operations, thus helping in reducing emissions of greenhouse gases and other harmful pollutants. Amongst such organizations are signatories to the United Nations' *Framework Convention on Climate Change* (UNFCCC) (1992) and other UN bodies that have, over the years supported such policies; like the *Paris Agreement* (2016), the *Sustainable Development Goals* (SDGs) (2015), the *New Urban Agenda* (NUA) (2016) and others that when implemented are aimed at helping to slow and curb the impacts of climate change. Nevertheless, despite a majority of member parties and signatories to these organizations and charters/agreements, it has not been an easy journey but one tediously slow (Monbiot, 2018; Plumer & Popovich, 2018). The hurdles in implementing these charters/agreements are borne by high-polluting countries whom perceive that ratifying these agreements will result in derailing their development agendas. Hence, such charters/agreements are an external threat to their economic prosperity aspirations.

Their political rhetoric coupled with convoluted implementation local bureaucracies seeks to skirt environmentally sensitive proposals sending dangerous political signals to competitors that they need to distance themselves from climate agreements for fear of their economic drawbacks. For instance, the United States' (US) decision to pull out of the *Paris Agreement* was not only based on the argument that the *Agreement* undermined the US economy, but also that similar negative effects would occur to the Chinese and Indian, with the former being its key economic and political rival (Rucker & Johnson, 2017). In reality, and with the urgency at which the issues of climate change need to be addressed, there are needs for concerted efforts from an array of players at various scales. Nevertheless, if that is the case, as will be demonstrated in the succeeding sections, economies—especially those in the Global South—will continue to reel and wallow in despair as the consequences of climate change catch up with them.

Cities and Climate Policy

In all the developmental transformations that the world has experienced over recent years—from agrarian, to industrialization and the technological revolution—cities were and still are always seen to play critical roles. As advanced in different literatures from many cultures (Collier, Jones, & Spijkerman, 2018; Gill & Goh, 2010; Mitra & Mehta, 2011; OECD,

2014), it is within cities that waves of development have been historically incubated, born and advanced to their maturity. For instance, during the industrialization era, major manufacturing innovations and transformations happened in cities. Similarly, as the world has shifted to a service industry base, being the predominant industry within cities today, cities have played the key facilitation role especially due to the establishment of critical infrastructures to support and enable large and growing populations while allowing the prospering of physical and digital consumerism (Turok & McGranahan, 2013) (Allam, 2020a, 2020b, 2020c, 2020d, 2020h; Allam, Tegally, & Thondoo, 2019). While there is no contest about the role of cities in sustaining our lifestyle today, it is additionally clearly apparent that cities have also been the leading contributors of harmful emissions during these various global developmental transformations. As an example, manufacturing industries trigger obvious large amounts of harmful surface, hydraulic and aerial emissions, thereby being major contributors of negative impacts to our environments at the same time as strengthening some economies—even at the expense of others. Today, these trends have not changed, and the consequences of urban activity upon our environments have worsened as we face unprecedented rates of emissions (IRENA, 2019; Kelsey & Meckling, 2018; Zhao & Zhang, 2018).

Since the last century, the world has experienced rapid population growth and urbanization, and these two phenomena have sparked demand for increased built environments, energy, food and water supplies, health and other resources. In addition, it is also in this period that the disposable income per capita has been seen to increase. This trend thus gives a sizeable number of people higher purchasing power. For this reason, global human consumption behaviour is argued to have changed resulting in higher demands for refined goods. Such demands in turn have been serving as catalyses for our manufacturing sectors to increase their production, and companies and firms have been nimbly on their toes to try to satisfy those increasing demands at their individual economic gains (Covert, Greenstone, & Knittel, 2016).

The pressure has also been on the construction industry where according to the United Nations, by 2100, the world will require at least 2 billion additional homes to accommodate the surging population (Smith, 2018). Such a huge demand will require massive consumptions of resources and spaces, while generating increased waste and emissions. In summary, such demands compound and confidently promulgate an overconsumption of resources, increased waste generation and increased unhealthy competition in the manufacturing sector and many other sectors. Unfortunately, since trends in population growth and urbanization do not show any signs of slowing down, at least until the year 2100 (Cilluffo & Ruiz, 2019), the urgency to devise alternative ways of satisfying global demands is real and urgent.

The above challenges have persisted amidst global calls for lasting solutions for issues of climate change to be pursued. One such call relates to the exploration of better ways towards streamlining the utilization of urban resources in ways that foster sustainability. The need for best practices to be adopted in urban resource consumption activities is prompted by the realization that cities contribute between 30 and 40% towards climate change (Satterthwaite, 2008), and that such statistics can be reduced if climate change mitigation policies are formulated and adopted especially in regard to urban design strategies. As an example, Reckien et al. (2018) demonstrate that such explorations are mostly sought by larger cities that accommodate a population of 500,000 or more people, with both mitigation and adaptability strategies being pursued. In the study, that entailed an examination of how 885 urban areas within 28 European Union (EU) countries had responded to the *Paris Agreement* concluding that it was clear that there is a long way to go before climate change policies are successfully enacted. Despite the EU being one of the major greenhouse gas emitters, only 26 and 17% of these 885 urban areas had invested in climate change mitigation and adaptation plans respectfully. Surprisingly, 33% of these urban areas were found to have no local climate change mitigation or adaptation plans.

As is posited by (Doherty, Klima, & Hellmann, 2016), the fight against climate change can only be won when the mitigation, adaptation and resilience agendas are initiated based upon local contextual needs. This argument is based upon recognizing the uniqueness's of each urban area in terms of their availability and consumption of resources and on how much emissions those contribute, along with the type of investment each has put forth to provide reduce those emissions. This is credited to locality-specific actions as the most potent change option because each city and urban area has different innovation capacities (Nordgren, Stults, & Meerow, 2016). When these innovation capacities are tailored to address specific local climate change issues, they can be very effective. It is also argued that when initiatives to address climate change are initiated and adopted to address local issues first, it becomes easier to incorporate the inclusion and empowerment of local people and communities in these initiatives more often resulting in positive local climate change mitigation strategies (Lesnikowski, Ford, Biesbroek, Berrang-Ford, & Heymann, 2016). Therefore, in the long run, when such local strategies are compounded, they are found to have far-reaching global impacts in combating climate change, and larger organizations have often to have taken stock as a consequence of local momentums for fear of losing their catchments and market palatability.

The need for mitigation and adaptability strategies are emphasized in policy documents through the *SDG* 11 and the *New Urban Agenda* (United Nations, 2016) brought forth by the UN. As noted above, these two policy documents emphasize the need for local-based solutions for tackling major urban challenges that derail the achievement of resilient societies and economies and thereby induce urban-related actions that exacerbate climate change. By adhering to the calls and strategies promoted in these documents, and other regional policies such as the *Paris Agreement* (2015) and the *Katowice Climate Change Conference Package* (2018), the signatory nations are bound to enact better resilience measures in regard to climate change towards increasing liveability and health in cities. As promoted in the SDG 11, there would be a reduction in social and economic inequality which in turn would promote conscious action towards sustainability (Marsal-Llacuna, Colomer-Llinàs, & Meléndez-Frigola, 2015). Similarly, as per the *Katowice Climate Change Conference Package*, urban areas, particularly in developing countries that would somehow struggle to catch up to the mitigation and adaptability strategy, would benefit from financial support from financing kits earmarked for the purposes the *Paris Agreement* (UNFCCC, 2018). This way, the contribution of cities on climate change can be lowered while at the same time while aiming to achieve increased liveability and resiliency. As further discussed in the succeeding section, locally initiated climate mitigation innovation strategies do not only benefit the local economy, but often impact positively upon global climate change actions.

Urban Economic Loss from Climate Change

The amount of time spent in different forums discussing and strategizing on how to combat climate change is a testament that indeed, this is a global human challenge that cannot be overlooked due to the increased pressures by various groups, and recently during the 'People's Climate March' in Washington, USA (Figs. 1.5, 1.6, and 1.7).

For this reason, efforts, commitments and actions towards achieving the agreed proposal, more so the *Paris Agreement*, should not be relegated to a level where the political class and the elite have a final say without regard to what the general populace think is right. The backdrop of this argument is based on an action taken by the US, through President Trump's withdrawing from the *Paris Agreement* (Leggett, 2019). What was disturbing in this action was labelling the issue of climate change 'a hoax'. The allegation disregarded the numerous, tangible challenges that the world is facing courtesy of this global challenge. Though Trump's actions were seen as reactionary, since the *Paris Agreement* is deemed by some governments in developing countries as possessing strategies to discourage or disrupt their respective economic agendas, his utterances stirred some level of confusion and prompted other heavy polluter nations to follow suite thereby jeopardizing all the efforts and gains made in the journey till now.

Fig. 1.5 People's Climate March (Image by Vlad Tchompalov)

Fig. 1.6 People's Climate March (Image by Vlad Tchompalov)

Fig. 1.7 People's Climate March (Image by Vlad Tchompalov)

Despite the positive views of developing countries on climate change relative to their spirited geopolitical economic agenda, it is very evident that climate change is very real. Of this evidence, based upon an analysis of data from a real-time perspective, it is clear that global temperatures have been affected, and reports from different parts of the world confirm this (IEA, 2017; IEA & UNEP, 2018; IRENA, IEA, & REN21, 2018). It is equally clear that local communities are now experiencing increasing threats from disruptive weather patterns and incidents. Chow (2018) highlights that incidents of extreme heat waves and cold snaps are now on the rise in different parts of the world, especially in coastal cities. In a report by the IPCC (IEA, 2017), global temperatures have been documented as rising by approximately 0.8–1.2 °C. As of 2018, it is reported that human-related activities have prompted approximately a 1.0 °C rise in temperatures above pre-industrial levels (temperatures between 1850 and 1900 were averaged at about 0.8 °C) (Hawkins et al., 2017). According to this report, if these trends continue unabated, between 2030 and 2052, global warming would rise to 1.5 °C above pre-industrial levels (IEA, 2017). Parallel to these increases in global temperatures, it is reported that global sea levels have increased approximately by 1.7 cm following the melting of glaciers. Inland, such increased water levels have resulted in increased flooding occurrences, and increased precipitation falls and more powerful hurricanes, typhoons and cyclones, just to name a few (Tagg, Räikkönen, Mäki, & Roca, 2016).

The impacts of extreme weather occurrences are significantly experienced in urban areas. This is because, as they have especially in the recent past, have had devastating effects upon critical city infrastructures. For instance, in 2015, it is reported that the Pilgrim Nuclear Power Station in Massachusetts had its power disrupted following increased seawater temperatures that served as the station's primary influent (Abel, 2015). In the same year, Räikkönen et al. (2016) documented how excessive wet snow affected power distribution in the region of Pirkanmaa in Finland causing a massive power outage that left thousands in darkness for hours. In 2019, following the destructive Cyclone Idai, it is reported that, besides the massive physical destruction it left in its trail, the city of Beira in Mozambique experienced over 90% destruction including its critical infrastructures like dams, roads, bridges and power distribution infrastructure falling victim (Onishi & Moyo, 2019). In addition to the physical destruction of infrastructures and other properties, there have been numerous cases of human injuries, displacements and loss of human lives due to climate change in different parts of the globe and little records about the non-human loss of life that would have equally have been numerically frightening.

According to a report by the UN Office for Disaster Risk Reduction, although there were over 10,373 people who lost their lives from earthquakes and tsunamis in 2018, weather-related disasters and incidents are causing much more devastating impacts, affecting over 61.7 million people in one way or another. This report indicates that in the period between 2000 and 2017, there was an average of 77,144 deaths recorded due to extreme weather catastrophes totalling to 1,221,465 reported human deaths between 2000 and 2018 (UNDRR, 2019).

In terms of urban fabric, the impacts of climate change have been immense with far-reaching consequences upon the social, political and economic dimensions of the world's cities. In respect to social aspects, the disruption of critical infrastructures has resulted in a myriad of challenges ranging from reduced supply of basic resources including food, water, energy and construction materials (Moazami, Nik, Carlucci, & Geving, 2019). It is evident that any form of disruption upon critical urban infrastructures impedes essential service deliveries like evacuation of people in risky grounds, slow supply of basic resources like foods and medications, and an increase in burdens such as housing. Such challenges are amplified even more when the infrastructure affected happens to be related to power generation or distribution. These trigger cascading and spiral effects, where other infrastructures though not directly affected, also ends up collapsing as a majority of infrastructure rely upon power to run effectively. For instance, in the Italian Dolomites, it was reported that in 2017 a major power blackout orchestrated by the disruption of electricity distribution by extreme ice concentrations left over 400,000 households in darkness and with dysfunctional amenities for three days (INMR, 2019).

On the economic front, disruptions or destructions of urban infrastructure impact not only affected areas but also entire national economies, and to some extent, regional and global economies. A case in point is the devastating impact of category 5 Hurricane Katrina on the US economy, where the accumulated loss was estimated to be over USD$100 billion and resources worth over USD$200 billion were used to reinstate cities back to operational capacities (Deryugina, Kawano, & Levitt, 2018; Plyer, 2016). In 2014, another extreme weather-induced catastrophe through the severe Typhoon Rammasun (Glenda) caused an overall economic loss of over USD$8.08 billion in the Philippines, Vietnam, China and Guam, while leaving a human death toll of 222 (CDRC, 2014). In addition to the destruction of infrastructures, as demonstrated here, climate change also prompts massive disruption to the proper functioning of urban fabrics and has overall negative consequences upon liveability, resiliency and sustainability. Thus, stern actions that are independent from political perspectives and values need to be taken.

The Need for Urban Resilience

Due to the eminent and increased impacts on human liveability, and the health and economic structures of cities, there have been spirited efforts to address the challenges of climate change. A bigger portion of the previous sections above has concentrated upon international strategies and policies that many governments have adopted. But as noted in the current section, due to contextual challenges, nationally specific policies are equally important in providing tailored responses to both local and global challenges.

As demonstrated in the NewClimate Policy Database (2015), the specificity of these policies is guided by issues including energy demand which varies in different economies. Also, demands and consumptions in areas like transportation and communication, the construction and building sector and in the industries sectors are argued to prompt the need for specific national policies (Schmidt & Fleig, 2018). Ultimately, the objective of these national policies, besides meeting their international obligations, is to assist local governments to conserve the wide array of their urban assets that plays a critical role in sustaining economic growth.

Iacubuta, Dubash, Upadhyaya, Deribe, and Höhne (2018) support this argument concluding that different countries differ in terms of urbanization trends, population increases and levels of income. These trends and increases are said to influence issues of climate change differently. Thus, they require governments to formulate specific policies that fit their national challenges and values. For instance, a quick look at the types of climate policies that countries in Europe have adopted vary greatly as similarly to those adopted in African countries or Asian countries. On this, South Africa is the only African country with the highest number of renewable energy policies totalling 25, while in Asia, Japan, India, China and others have over 50 renewable energy policies. In the European Union (EU), there are over 80 policies on renewable energy, over 80 on nuclear energy and over 100 on energy efficiency (NewClimate

Policy Database, 2015). There is also a study by Ylä-Anttila et al. (2018) that highlights that ambitions to enact climate change policies by different countries are also influenced by their political structures and their national culture of science. They conclude that economic activities and human decisions to enact environmental policies are hampered by the national political leadership of the day, and these political regimes directly influence policies drafted by formulating bodies in each country.

Despite the uniqueness of different countries, global national policies are guided by different policy instruments proposed by the *IEA/IRENA Global Renewable Energy Policies and Measures Database*, and these include categorizations including economic instruments, regulatory instruments, information and education, policy support, research and development (R&D) and voluntary approaches, barrier removal, and climate strategies and targets. This is why such national policies align well with the international policies, and in most case, most national policies are prompted by international climate policies. For instance, Iacubuta et al. (2018) note that after the *Paris Agreement* in 2015, the number of domestic greenhouse gas (GHG)-oriented policies globally increased from 48% as recorded in 2012 to approximately 89% by 2017. Höhne et al. (2017) explain that national policies came about as member countries of the United Nations *Framework Convention on Climate Change* (UNFCCC), party to the *Paris Agreement*, endeavoured to put in place measures that could allow them to remain faithful to this *Agreement* of which 55 member parties had ratified by the third day whereupon the *Agreement* came into force on 4 November 2016. To date, 185 out of 197 party members have ratified the document (UNFCCC, 2019).

Besides policies touching on global warming, agricultural policies that emphasize a raft of measures to be undertaken and implemented to increase food security have also gained popularity in many countries internationally. This acceptance follows several severe droughts over this decade that have contributed to loss of human lives in different parts of the world. For instance, according to a report by World Vision, more than 285,000 people lost their lives between 2011 and 2012 in East Africa alone due to this crisis (Vision, 2019). Beside this death toll, according to the Food and Agricultural Organization (FAO), in 2016 alone, 815 million people globally were experiencing chronic undernourishment, and this situation continues to worsen at an alarming rate (FAO, IFAD, UNICEF, WFP, & WHO, 2017). Unfortunately, these and a myriad of other new climate change-related policies formulated after various disasters have occurred cannot remediate past lost lives and landscapes. But they are going towards playing a significant role in helping to negate and or mediate further loss of human life and productive landscapes. Nevertheless, governments need to be proactive, and they can, and where possible, to ensure that they take advantage of modern prediction and modelling tools so they can plan beforehand.

In addition to the formulation of new policies, calls for resilience planning are also being heeded through the numerous urban regeneration initiatives that are taking place in different cities across the world. Strategies like, making best use of redundant brownfields or capitalizing on underutilized spaces by concentrating and actualizing new development in them are common strategies to help curb urban sprawl. The adoption of mixed-use development and compact neighbourhoods, like in case of Singapore (Chia, Li, & Yang, 2017), and Kanazawa City in Japan (Balaban & Puppim de Oliveira, 2013), are a few examples. Urban regeneration geared towards climate change mitigation is also seen to focus upon addressing issues like transportation and mobility, energy consumption and efficiency, and the optimization of infrastructure for enhanced resource efficiencies, to name a few outcomes (Balaban & Puppim de Oliveira, 2013). The regeneration of existing built environment is perceived as a potent strategy to help achieve increased efficiencies and economic performances while addressing the liveability dimensions of a city without compromising on resilience.

Green Funding, a Dry-Cleaning Story

Due to the numerous and dire consequences prompted by climate change inspired events as discussed above, the world has experienced substantial, concerted efforts and calls from interdisciplinary agencies all geared towards finding appropriate and lasting climate change mitigation tools and initiatives. As noted previously, numerous conventions have been held and agreements reached on how the world needs to move forward with immediate and future actions to ensure that harmful pollutants that accentuate climate change are reduced. In these deliberations, most of the proposed strategies involve substantial investment in hard infrastructures which have the capacity to help in mitigation and adaptability (Ehlers, 2014; Engel, Fischer, & Galetovic, 2010). Examples of such infrastructures include renewable energy power plants, and water reservoirs and dams especially in dry and semi-arid areas to ensure sufficient secure supplies of water for domestic and agricultural activities. Such also includes infrastructures that promote food productivity, food security and waste management, etc. Unfortunately, investing in these infrastructures is an incredibly expensive affair that only a few economies can comfortably manage without having to sources and secure major internal or external loans, or without seeking financial assistance from green climate fund sources. Such financial requirements have been seen to discourage countries, especially those in the Global South from investing in substantial critical infrastructures despite being on the front line of climate change (Colenbrander, Lindfield, Lufkin, & Quijano, 2018; Kellett & Caravani, 2013).

Challenges in theses investment also arise from the truth that special green climate funds are seldom sufficient, and there is an array of additional requirements that a country is required to meet before it can get access to those funds (Linnér & Klein, 2017). It is argued that many governments seeking these climate funds are often reluctant to adhere to prudent financial practices, and a majority of these governments are accused of requiring the funds to be replenished without anything tangible to show for previous disbursements. These funds also face unprecedented politics and tension especially between rich and poor countries due to the nature of how membership structures for global green funds are configured. Ideally, every country ought to have an equal say on how the fund is to be distributed and advanced to the most deserving. But, in reality there are numerous notable wrangles within the structure as representatives of poor and rich countries do not agree on matters and priorities as to how finances should be raised, utilized and distributed (Darby & Mathiesen, 2018). These challenges, around green funds, appear somehow unattractive especially to poor countries, though they need these funds if they are to secure their economies and societies from the harsh realities of climate change.

Besides inequalities in the distribution of financial support to invest in new, or in the upgrading of existing infrastructures, there is also an environment of unfairness in the implementation of the environmental policies as regions who pollute less are seen to be suffering the most in deference to those who pollute more (Plumer & Popovich, 2018). In some quarters, the actions of high-polluting countries are seen as deliberate due to their inconsiderate linear economic agendas even though research upon climate change impacts are well established and known not to be localized and would have far-reaching consequences globally (Cole, 2015; Rao, 2014). With this knowledge, and due to their economic capabilities, these countries have invested heavily in infrastructural developments that shield them from the most of the consequences of climate change but conversely their counterparts in low-emitting countries have serious infrastructural deficits thereby rendering them vulnerable to climate change (Khaqqi, Sikorski, Hadinoto, & Kraft, 2018). As evidence of this pattern, Althor, Watson, and Fuller (2016) found out that 20 out of 36 countries, with the highest emitting capacities, are classified amongst those with the least vulnerabilities from climate change, while 11 out of 17 with least emissions are classified in the list of those with high risks of vulnerability. This unfairness is further highlighted by statements from international climate change agencies, and this is why policies proposed in conventions by UNFCCC that highly advocate for common responsibility in reducing emissions commensurate to emissions capacities. Ideally, this means that those polluting more need to likewise reduce more, but in reality, this has not been the case. Thus, when high polluters perceived that they are being pushed to the limit, they threaten to withdraw altogether from their global obligations. This threat does not only mean that they would continue emitting without care for what environmental policies dictates, but they also withdraw their financial and non-financial contributions to various climate funds, and this has a flow-on effect in severely impacting on global green fund availability as most of these countries contribute significant amounts to this fund. A case in point is the announced withdrawal from the *Paris Agreement* by the US which also means termination of their financial obligations and donations towards environmental financing (Zhang, Chao, Zheng, & Huang, 2017).

Despite this, even with contributions of green funds by these high polluters such investments are sometimes contestable as they are seen to be making these contributions as a face-saving strategy from adverse media attention and negative branding they receive due to their blatant continued emission releases despite global calls for reduction.

The sustainability of green funds is thus an interesting case because the source of these funds, as explained above, mainly originates from economies that more often continue to engage in unsustainable and high emissions practices. Yet, given the urgency of climate change, the global debate on the ethics and morality of the origins of such funds must not be a deterrent for their use, because those funds have significantly helped tangible infrastructural development in poor and vulnerable economies.

Hence, there is a need to ensure sufficient equitable fund access to protect and assist vulnerable communities (S. Spratt & Griffith-Jones, 2013). The world overall has a moral responsibility to not remain voiceless, static and helplessly watch developed countries continue emitting unabated pollutants simply because they are contributing to green funds. After all, it is evident that these funds may not be adequate in achieving long-term and viable solutions to tackling global climate change. This is especially pertinent in terms of their contestable sustainability and the challenges that crowd access to funds. Also, as emissions continue to increase, costs continue to rise, the funds may never be enough to cover all the needy, global cases that urgently need to be addressed. In essence, if emission reduction policies are made to work effectively, and polluters are made to compensate on their ills, there will be a reduction in emissions thereby lessening the urgency to invest in critical, climate change mitigation infrastructure resulting possibly in more manageable and even perhaps neutralized climate change emissions. To align with this numerous mechanisms working towards the reduction of transaction costs can also effectively aid in accelerating the adoption of green mitigation tools (Shahab, Clinch, & O'Neill, 2017, 2018a, 2018b; Shahab & Viallon, 2019).

Climate Technology and Fortune Telling

Interestingly, and thankfully, the challenges of climate change have prompted further actions beyond the direct climate change mitigation approaches. There is now increased international emphasis upon investment in Research and Development (R&D) for climate prediction tools geared towards making cities more resilient to varying weather impacts. The motivation here is based on the evidence that contemporary understandings of historical trends of certain climatic conditions and how such have been impacted by emissions can assist in predicting how cities may be impacted in the future (Bibri, 2018; Kumar, Singh, Ghosh, & Anand, 2012). Dixon, Montgomery, Horton-Baker, and Farrelly (2018) note that since cities were built to last hundreds or thousands of years, their planning today needs to be sensible to better address challenges from both populations increase and even more sensitively and nimbly the impacts of climate change. Such factors prompt the need for proper and well-informed risk mitigation and adaptation plans that can only be achieved when there is spirited investment in R&D. For this reason, there has now been a shift in interest in engaging various disciplines, including soft computing and artificial intelligence (AI) in predicting weather patterns with more precision enabling climate change mitigation solutions to be nimbly calibrated to cater from current and future needs (Li, Xu, Hu, Xiao, & Wang, 2018; Namitha, Jayapriya, & Kumar, 2015; Yer, 2017). The results from the use of these technologies are said to be spurring new digital innovations and knowledge, more so in areas that ultimately can lead to reductions in emissions and the associated costs of mitigation measures.

From discussions in the previous sections, it was pointed out that only 17% of 885 urban areas in 28 European countries had invested and implemented adaptation plans. One reason for this paucity of investment and acceptance is the high financial demand required to invest in climate change adaptable infrastructures. The other reason is simply ignorance from both local and urban leaders on what the future holds to their locale and city landscapes when it comes to climate change. Immediacy and perceived slow timelines in witnessing tangible impacts compound the latter. Therefore, with the advent of new cutting-edge technologies and tools, that have the capacity to more accurately predict, model and scenario-plan how a city would be impacted in circumstances of stress, we are increasingly better positioned to orally and visually communicate to leaders and communities about these complex matters and the need for tangible actions to occur in timely addressing adaptation strategies. Moreover, it is argued that with statistically valid and visually clear information on how certain climatic events would unfold in the coming years for a city and their associated landscape, cities are more likely to increase their budgetary provisions in investing in better adaptation and resilience projects from the current meagre 2–5% to a % figure more commensurate of their risk (A. Chang, 2018). On this, as advanced in an OECD (2008) report, the results obtained from prediction tools can help to document and validate the need for sufficient budgetary allocations that can be set aside to facilitate mitigation projects. The more the available statistical and visual evidence, the better-informed cities are in moving closer to achieving adaptability and resiliency.

Today, predictions articulated from different agencies and scientific bodies evidence that the world is environmentally nose-diving into an era with dire consequences arising from climate change, and the impacts of this pattern are having a direct negative and devastating effect upon the world's ecosystems, animal communities and upon human liveability generally. Of the many consequences, the most astonishing prediction is that global temperatures are likely to rise by approximately 4 °C if the current patterns of emissions and global warming events continue. The OECD (2008) report argues that such startling temperature rises would only be prevented if numerous climate negotiations are put in practice within the earliest opportunity, but not later than 2020, with the proviso that there is no backtracking or retrospectively on adopted policies. Similar conclusions have been reached by D. Spratt and Dunlop (2019) who argue that the future of the globe hangs on how quickly a zero-emission industrial system can be assembled and implemented to replace what the world has now.

The above reports are just a sample of the numerous concerns and warnings that have been raised by a broader, concerned audience, who wants to see a major re-calibration of national and international priorities touching on climate change efforts. Today, secondary students, journalists, popular writers, media and visual image-makers are just a few of the groups that are joining in efforts to articulate and advocate their values on the need for a better and cleaner future where climate change will not be a problem. Streeby (2018) explains how such different groups are using their trade and bloc voices to (re)imagine how the world will be in the future if climate change continues unabated. There now numerous sci-fi books, movies and articles that highlight some of the narratives that would befall the earth and its occupants, including its flora and fauna in the future, in terrestrial, aquatic and aerial environments where things will have gone 'haywire' due to tangible and subtle climate change disasters, incidents and patterns (Kraus, 2017). On the argument, although mainstream media is often controlled and financed by polluting groups or their host country regimes, there is a dearth of journalists and media working towards sharing statistical and visual evidence about the need for amplifying speedy actions (Boykoff et al., 2018). The purpose of the latter is to evoke concern amongst all stakeholders, including citizens, to adopt sustainable practices. In contrast, others, like technology-oriented corporations are said to be spending huge amounts of money to invest in programmes that would help the general public remain committed to sustainability, or their version or narrative of sustainability, more so in terms of the utilization and commoditization of resources. On this, Microsoft, for instance, is said to engage in a project called '*AI for Earth project*' that entails designing and making smart gadgets and devices that have the potential to maximize energy use and thus reducing over-consumption (Jones, 2018), or are they inventing new consumption items geared towards sustainability receptive purchasers. Others like IBM and Cisco have also been heavily linked with actualizing smart cities concepts, geared towards making sure that urban areas have the potential and capacity to reduce their emissions. These corporations are rhetorically stating that their aims are to use technologies, such as big data, deep learning and AI, to streamline different urban fabrics to make them more adaptable and thus strengthening city sustainability.

The increasing public and corporate attention upon climate change have prompted even further steps where there are now policy proposals to integrate climate change content in primary, secondary and tertiary school curricula. This curricula policy shift is more evident in the built environment, engineering, applied science and information technology disciplines creating informed graduates and new market forces that are better equipped to innovatively cater for eminent climate change challenges. For example, Italy has just announced, in November 2019, the introduction of a mandatory climate change secondary school subject to be implemented across Italy in the shadow of their droughts and the sinking of Venice (Horowitz, 2019). In some quarters, such progressive proposals have already been actualized and evidenced in C.-H. Chang and Pascua (2017), whereby climate change education has in turn helped to shift the ethics of professional practice where even the young are now actively involved in articulating and practising climate change initiatives and are also advocating for the adoption of best practices. A case in point is the emergence of secondary student-led movements such as the *School Strike 4 Climate Action* whose main agenda is to persuade adults to take urgent climate actions (Aljazeera, 2019).

Amid all these concerted efforts of 'calls for action', as noted in the beginning of this section, the solution may lie in the utilization of cutting-edge climate change digital technologies to precisely predict the future to enable 'on-point' actions to be taken. For this reason, investing in R&D for climate purposes is no longer an option but is now a necessity to allow actual and unbiased predictions.

References

Abel, D. (2015, August 11). *Pilgrim facility cuts its power*. Retrieved from https://www.bostonglobe.com/metro/2015/08/11/high-water-temperatures-forced-power-cut-pilgrim-nuclear-plant/fMgG6VtRmadnVcuacbPpGI/story.html.

Aljazeera. (2019, May 24). *School students strike worldwide, demand action on climate change.* Retrieved from https://www.aljazeera.com/news/2019/05/school-students-strike-worldwide-demand-action-climate-change-190524192710600.html.

Allam, Z. (2012). Sustainable architecture: Utopia or feasible reality? *Journal of Biourbanism, 2,* 47–61.

Allam, Z. (2017). Building a conceptual framework for smarting an existing city in Mauritius: The case of Port Louis. *Journal of Biourbanism, 4*(1 & 2), 103–121.

Allam, Z. (2019a). The city of the living or the dead: On the ethics and morality of land use for graveyards in a rapidly urbanised world. *Land Use Policy, 87,* 104037.

Allam, Z. (2019b). Identified priorities for smart urban regeneration: Focus group findings from the city of Port Louis, Mauritius. *Journal of Urban Regeneration & Renewal, 12*(4), 376–389.

Allam, Z. (2020a). Data as the new driving gears of urbanization. In Z. Allam (Ed.), *Cities and the digital revolution: Aligning technology and humanity* (pp. 1–29). Cham: Springer.

Allam, Z. (2020b). Digital urban networks and social media. In Z. Allam (Ed.), *Cities and the digital revolution: Aligning technology and humanity* (pp. 61–83). Cham: Springer.

Allam, Z. (2020c). On culture, technology and global cities. In Z. Allam (Ed.), *Cities and the digital revolution: Aligning technology and humanity* (pp. 107–124). Cham: Springer.

Allam, Z. (2020d). Privatization and privacy in the digital city. In Z. Allam (Ed.), *Cities and the digital revolution: Aligning technology and humanity* (pp. 85–106). Cham: Springer.

Allam, Z. (2020e). Religion and urban planning. In Z. Allam (Ed.), *Theology and urban sustainability* (pp. 1–19). Cham: Springer.

Allam, Z. (2020f). Religious matrimony, urban sprawl and urban morphology. In Z. Allam (Ed.), *Theology and urban sustainability* (pp. 21–35). Cham: Springer.

Allam, Z. (2020g). Theology, sustainability and big data. In Z. Allam (Ed.), *Theology and urban sustainability* (pp. 53–67). Cham: Springer.

Allam, Z. (2020h). Urban chaos and the AI Messiah. In Z. Allam (Ed.), *Cities and the digital revolution: Aligning technology and humanity* (pp. 31–60). Cham: Springer.

Allam, Z., Dhunny, A., Siew, G., & Jones, D. (2018). Towards smart urban regeneration: Findings of an urban footprint survey in Port Louis. *Mauritius. Smart Cities, 1*(1), 121–133.

Allam, Z., & Jones, D. (2018). Towards a circular economy: A case study of waste conversion into housing units in Cotonou. *Benin. Urban Science, 2*(4), 118.

Allam, Z., & Jones, D. (2019). Climate change and economic resilience through urban and cultural heritage: The case of emerging small island developing states economies. *Economies, 7*(2), 62.

Allam, Z., Tegally, H., & Thondoo, M. (2019). Redefining the use of big data in urban health for increased liveability in smart cities. *Smart Cities, 2*(2), 259–268.

Althor, G., Watson, J. E. M., & Fuller, R. A. (2016). Global mismatch between greenhouse gas emissions and the burden of climate change. *Scientific Reports, 6*(20281), 1–6.

Apreda, C., D'Ambrosio, V., & Di Martino, F. (2019). A climate vulnerability and impact assessment model for complex urban systems. *Environmental Science & Policy, 93,* 11–26. https://doi.org/10.1016/j.envsci.2018.12.016.

Arnone, E., Pumo, D., Francipane, A., Loggia, G. L., & Noto, L. V. (2018). The role of urban growth, climate change, and their interplay in altering runoff extremes. *Hydrological Processes, 32*(12), 1755–1770.

Balaban, O., & Puppim de Oliveira, J. A. (2013). Understanding the links between urban regeration and climate-friendly urban development: Lessons from two case studies in Japan. *Local Environment, 19*(8), 869–890.

BBC News. (2018, September 5). *Typhoon Jebi forces closure of Kansai airport, near Osaka in Japan.* Retrieved from https://www.bbc.com/news/world-asia-45417035.

Bibri, S. E. (2018). The IoT for smart sustainable cities of the future: An analytical framework for sensor-based big data applications for environmental sustainability. *Sustainable Cities and Society, 38,* 230–253.

Boykoff, M., Andrews, K., Daly, M., Katzung, J., Luedecke, G., Maldonado, C., & Nacu-Schmidt, A. (2018). *A Review of Media Coverage of Climate Change and Global Warming in 2017.* Retrieved from Colorado: http://sciencepolicy.colorado.edu/icecaps/research/media_coverage/summaries/special_issue_2017.html.

CDRC. (2014, July 18). *Typhoon Glenda leaves more than 800,000 people affected.* Retrieved from http://www.cdrc-phil.com/typhoon-glenda-leaves-more-than-800000-affected/.

Chang, A. (2018, October 22). *Cities are planning for climate change all wrong.* Retrieved from https://www.washingtonpost.com/news/theworldpost/wp/2018/10/22/climate-change-2/.

Chang, C.-H., & Pascua, L. (2017). The state of climate change education—Reflections from a selection of studies around world. *International Research in Geoprahical and Environmental Education, 26*(3), 177–179.

Chia, W.-M., Li, M., & Yang, T. (2017). Public and private housing markets dynamics in Singapore: The role of fundamentals. *Journal of Housing Economics, 36,* 44–61.

Chow, W. T. L. (2018). The impact of weather extremes on urban resilience to hydro-climate hazards: A Singapore case study. *International Journal of Water Resources Development, 34*(4), 510–524.

Cilluffo, A., & Ruiz, N. G. (2019, June 17). *World's population is projected to nearly stop growing by the end of the century.* Retrieved from https://www.pewresearch.org/fact-tank/2019/06/17/worlds-population-is-projected-to-nearly-stop-growing-by-the-end-of-the-century/.

Cole, D. H. (2015). Advantages of a polycentric approach to climate change policy. *Nature Climate Change, 5,* 114–118.

Colenbrander, S., Lindfield, M., Lufkin, J., & Quijano, N. (2018). *Financing low-carbon, climate-resilient cities.* London and Washington, DC: Coalition for Urban Trasitions.

Collier, P., Jones, P., & Spijkerman, D. (2018). *Cities are engines of growth: Evidence from a new global samples of cities.* Retrieved from http://pubdocs.worldbank.org/en/356001527999188468/text/117-Cities-as-Engines-of-Growth-Collier-et-al-28March2018.txt.

Condon, P. M., Cavens, D., & Miller, N. (2009). *Urban planning tools for climate change mitigation.* Cambridge, MA: Lincoln Institute of Land Policy.

Corey, K. E., Wilson, M. I., & Fan, P. (2015). Cities, technologies and economic change. In R. Paddison & T. Hutton (Eds.), *Cities and economic change: Restructuring and dislocation in the global metropolis.* Thousand Oaks, CA: Sage.

Covert, T., Greenstone, M., & Knittel, C. R. (2016). Will we ever stop using fossil fuels? *The Journal of Economic Perspectives, 30*(1), 117–137.

Darby, M., & Mathiesen, K. (2018, August 9). *Crucial global climate fund facing massive challenges.* Retrieved from https://www.climatechangenews.com/2018/08/09/crucial-global-climate-fund-facing-massive-challenges/.

Deryugina, T., Kawano, L., & Levitt, S. (2018). The economic impact of hurricane Katrina on its victims: Evidence from individual tax returns. *American Economic Journal: Applied Economics, 10*(2), 202–233.

Dixon, T., Montgomery, J., Horton-Baker, N., & Farrelly, L. (2018). Using urban foresight techniques in city visioning: Lessons from the reading 2050 vision. *Local Economy, 33*(8), 1–23.

Doherty, M., Klima, K., & Hellmann, J. J. (2016). Climate change in the urban environment: Advancing, measuring and achieving resiliency. *Environmental Science & Policy, 66,* 310–313.

Ehlers, T. (2014). *Understanding the challenges for infrastructure finance* (BIS Working Papers 454). Bank for International Settlements.

Engel, E., Fischer, R., & Galetovic, A. (2010). The economics of infrastructure finance: Public private partnerships versus public provision. *EIB Papers, 15*(1), 40–69.

FAO, IFAD, UNICEF, WFP, & WHO. (2017). *The state of food security and nutrition in the world 2017: Building resilience for peace and food security.* Retrieved from Rome.

Ghebrezgabher, M. G., Yang, T., & Yang, X. (2016). Long-term trend of climate change and drought assessment in the Horn of Africa. *Advances in Meteorology, 2016*(2), 1–12.

Gill, I. S., & Goh, C.-C. (2010). Scale economies and cities. *The World Bank Research Observer, 25*(2), 235–262.

Hawkins, E., Ortega, P., Suckling, E., Schurer, A., Hegerl, G., Jones, P., et al. (2017). Estimating changes in global temperatures since the preindustrial period. *Bulletin of the American Meteological Society, 98*(9), 1841–1856.

Höhne, N., Kuramochi, T., Warnecke, C., Röser, F., Fekete, H., Hagemann, M., et al. (2017). The Paris Agreement: Resolving the inconsistency between global goals and national contributions. *Climate Policy, 17*(1), 16–32.

Horowitz, J. (2019). *Italy's students will get a lesson in climate change. Many lessons, in fact.* Retrieved from https://www.nytimes.com/2019/11/05/world/europe/italy-schools-climate-change.html.

Iacubuta, G., Dubash, N. K., Upadhyaya, P., Deribe, M., & Höhne, N. (2018). National climate change mitigation legislation, strategy and targets: A global update. *Climate Policy, 18*(9), 1114–1132.

IEA. (2017). *World energy outlook 2017.* Paris, France: OECD/IEA.

IEA, & UNEP. (2018). *2018 Global Status Report: Towards a zero-emission, efficient and resilient buildings and construction sector.* Retrieved from https://www.worldgbc.org/news-media/2018-global-status-report-towards-zero-emission-efficient-and-resilient-buildings-and.

INMR. (2019, July 13). *Impact of climate change on power systems & electrical insulation: Experience in Italy.* Retrieved from https://www.inmr.com/impact-climate-change-power-systems-electrical-insulation-experience-italy/.

IRENA. (2019). *Renewable energy market analysis: GCC 2019* (IRENA Ed.). Abu Dhabi.

IRENA, IEA, & REN21. (2018). *Renewable energy policies in a time of transition.* IRENA, OECD/IEA and REN21.

Jones, N. (2018). *Can artificial intelligence help build better, smarter climate models.* Retrieved from https://e360.yale.edu/features/can-artificial-intelligence-help-build-better-smarter-climate-models.

Kellett, J., & Caravani, A. (2013). *Financing disaster risk reduction: A 20 year story of international aid.* London: ODI.

Kelsey, N., & Meckling, J. (2018). Who wins in renewable energy? Evidence from Europe and the United States. *Energy Research & Social Science, 37,* 65–73.

Khaqqi, K. N., Sikorski, J. J., Hadinoto, K., & Kraft, M. (2018). Incorporating seller/buyer reputation-based system in blockchain—Enabled emission trading application. *Applied Energy, 209,* 8–19.

Kraus, R. (2017). *Hollywood is obsessed with climate change disasters. What does that mean for the planet?* Retrieved from https://www.thrillist.com/entertainment/nation/climate-change-disaster-movies.

Kumar, A., Singh, M. P., Ghosh, S., & Anand, A. (2012). Weather forecasting model using artificial neural network. *Procedia Technology, 4,* 311–318.

Lang, G. (2018). Urban energy futures: A comparative analysis. *European Journal of Futures Research, 6*(19), 1–19.

Leggett, J. (2019). *Potential implications of U.S. withdrawal from the Paris Agreement on climate change.* Retrieved from https://fas.org/sgp/crs/misc/IF10668.pdf.

Lesnikowski, A., Ford, J., Biesbroek, R., Berrang-Ford, L., & Heymann, J. (2016). National-level progress on adaptation. *Nature Climate Change, 6,* 261–264.

Li, H., Xu, X., Hu, Y., Xiao, Y., & Wang, Z. (2018). Assimilation of Doppler radar data and its impact on prediction of a heavy Meiyu frontal rainfall event. *Advance in Meteorology, 2018,* 1–14.

Linnér, B.-O., & Klein, R. J. T. (2017). Climate change adaptation and the least developed countries fund (LDCF): Qualitative insights from policy implementation in the Asia-Pacific. *Climatic Change, 140*(2), 209–226.

Marsal-Llacuna, M. L., Colomer-Llinàs, J., & Meléndez-Frigola, J. (2015). Lessons in urban monitoring taken from sustainable and livable cities to better address the smart cities initiative. *Technological Forecasting and Social Change, 90*(B), 611–622.

Mgbemene, C. A., Nnaji, C. C., & Nwozor, C. (2016). Industrialization and its backlash: Focus on climate change and its consequences. *Journal of Environmental Science and Technology, 9,* 301–316.

Mi, Z., Guan, D., Liu, Z., Viguie, V., Fromer, N., & Wang, Y. (2019). Cities: The core of climate change mitigation. *Journal of Cleaner Production, 207,* 582–589.

Miller, J. D., & Hutchins, M. (2017). The impacts of urbanisation and climate change on urban flooding and urban water quality: A review of the evidence concerning the United Kingdom. *Journal of Hydrology: Regional Studies, 12,* 345–362.

Mitra, A., & Mehta, B. (2011). Cities as the engine of growth: Evidence from India. *Journal of Urban Planning and Development, 137*(2), 171–183.

Moazami, A., Nik, V. M., Carlucci, S., & Geving, S. (2019). Impacts of future weather data typology on building energy performance—Investigating long-term patterns of climate change and extreme weather conditions. *Applied Energy, 238*, 696–720.

Monbiot, G. (2018, September 26). *While economic growth continues we'll never kick our fossil fuels habit*. Retrieved from https://www.theguardian.com/commentisfree/2018/sep/26/economic-growth-fossil-fuels-habit-oil-industry.

Namitha, K., Jayapriya, A., & Kumar, G. S. (2015). Rainfall prediction using artificial neural network on map-reduced framework. *Proceedings of the Third International Symposium on Women in Computing and Informatics—WCI.*

NewClimate Policy Database. (2015). *Database structure and data categorisation*. Retrieved from http://climatepolicydatabase.org/index.php?title=Database_structure_and_data_categorisation.

NOAA. (2018, June). *Hurricanes and tropical storms—Annual 2017*. Retrieved from https://www.ncdc.noaa.gov/sotc/tropical-cyclones/201713/.

NOAA. (2019). *Weather disasters and costs*. Retrieved from https://coast.noaa.gov/states/fast-facts/weather-disasters.html.

Nordgren, J., Stults, M., & Meerow, S. (2016). Supporting local climate change adaptation: Where we are and where we need to go. *Environmental Science & Policy, 66*, 344–352.

OECD. (2008). *Climate change mitigation: What do we do?* OECD.

OECD. (2014). *Cities and climate change: National governments enabling local action*. OECD Policy Perspectives. OECD.

Ogden, L. E. (2018). Climate change, pathogens, and people: The challenges of monitoring a moving target. *BioScience, 68*(10), 733–739.

Onishi, N., & Moyo, J. (2019, March 18). *Cyclone Idai destroys 'percent' of a city of half a million in Southern Africa*. Retrieved from https://www.nytimes.com/2019/03/18/world/africa/cyclone-idai-mozambique.html.

Plumer, B., & Popovich, N. (2018, December 7). *The world still isn't meeting its climate goals*. Retrieved from https://www.nytimes.com/interactive/2018/12/07/climate/world-emissions-paris-goals-not-on-track.html.

Plyer, A. (2016, August 26). *Facts for features: Katrina impact*. Retrieved from https://www.datacenterresearch.org/data-resources/katrina/facts-for-impact/.

Population Reference Bureau. (2018). *2018 world population data sheet*. Retrieved from Washington, DC.

Räikkönen, M., Mäki, K., Murtonen, M., Forssén, K., Tagg, A., Petiet, P. J., et al. (2016). A holistic approach for assessing impact of extreme weather on critical infrastructure. *International Journal of Safety and Security Engineering, 6*(2), 171–180.

Rao, N. D. (2014). International and intranational equity in sharing climate change mitigation burdens. *International Environmental Agreements: Politics, Law and Economics, 14*(2), 129–146.

Reckien, D., Salvia, M., Heidrich, O., Church, J. M., Pietrapertos, F., Gregorio-Hurtado, S. D., et al. (2018). How are cities planning to respond to climate change? Assessment of local climate plans from 885 cities in the EU-28. *Journal of Cleaner Production, 191,* 207–219.

Rezai, A., Taylor, L., & Foley, D. (2018). Economic growth, income distribution, and climate change. *Ecological Economics, 146,* 164–172.

Rucker, P., & Johnson, J. (2017, June 1). *Trump announces U.S. will exit Paris climate deal, sparking criticism at home and abroad.* Retrieved from https://www.washingtonpost.com/politics/trump-to-announce-us-will-exit-paris-climate-deal/2017/06/01/fbcb0196-46da-11e7-bcde-624ad94170ab_story.html?noredirect=on&utm_term=.1c83f7567408.

Satterthwaite, D. (2008). Cities' contribution to global warming: Notes on the allocation of greenhouse gas emissions. *Environment and Urbanization, 20*(2), 539–550.

Schmidt, N. M., & Fleig, A. (2018). Global patterns of national climate policies: Analyzing 171 country portfolios on climate policy integration. *Environmental Science & Policy, 84,* 177–185.

Shahab, S., Clinch, J. P., & O'Neill, E. (2017). Impact-based planning evaluation: Advancing normative criteria for policy analysis. *Environment and Planning B: Urban Analytics and City Science, 46*(3), 534–550.

Shahab, S., Clinch, J. P., & O'Neill, E. (2018a). Estimates of transaction costs in transfer of development rights programs. *Journal of the American Planning Association, 84*(1), 61–75.

Shahab, S., Clinch, J. P., & O'Neill, E. (2018b). Accounting for transaction costs in planning policy evaluation. *Land Use Policy, 70,* 263–272.

Shahab, S., & Viallon, F.-X. (2019). A transaction-cost analysis of Swiss land improvement syndicates. *Town Planning Review, 90*(5), 545–565.

Smith, S. (2018). *The world needs to build more than two billion new homes over the next 80 years.* Retrieved from https://theconversation.com/the-world-needs-to–more-than-two-billion-new-homes-over-the-next-80-years-91794.

Spratt, D., & Dunlop, I. (2019). *Existential climate-related security risk: A scenario approach.* Retrieved from https://docs.wixstatic.com/ugd/148cb0_b2c0c79dc4344b279bcf2365336ff23b.pdf.

Spratt, S., & Griffith-Jones, S. (2013). *Mobilising investment for inclusive green growth in low-income countries.* Retrieved from Eschborn, Germany.

Streeby, S. (2018). *Imagining the future of climate change.* Oakland: University of California Press.

Tagg, A., Räikkönen, M., Mäki, K., & Roca, M. (2016). Impact of extreme weather on critical infrastructure: the EU-INTACT risk framework. *E3S Web of Conferences, 7,* 07007.

Turok, I., & McGranahan, G. (2013). Urbanization and economic growth: The arguments and evidence for Africa and Asia. *Environment and Urbanization, 25*(2), 465–482.

UN-Habitat. (2015). *Urbanization and climate change in small island developing states.* Retrieved from Nairobi, Kenya: https://unhabitat.org/wpdm-package/urbanization-and-climate-change-in-small-island-developing-states/?wpdmdl=114762.

UNDRR. (2019, January 24). *2018: Extreme weather events affected 60m people.* Retrieved from unisdr.org/archive/63267.

UNFCCC. (2018). *COP 24 agenda as adopted.* Retrieved from https://unfccc.int/node/185180.

UNFCCC. (2019). *Paris Agreement—Status of ratification.* Retrieved from https://unfccc.int/process/the-paris-agreement/status-of-ratification.

United Nations. (2016). *The new urban agenda.* Paper presented at the United Nations Conference on Housing and Sustainable Urban Development (Habitat III), Quito, Ecuador.

United Nations. (2017). *World population prospects: Key findings & advance tables.* New York, NY: Department of Economic and Social Affairs, Population Division.

Vision, W. (2019). *East Africa hunger, famine: Facts, FAQs, andhow to help.* Retrieved from https://www.worldvision.org/hunger-news-stories/east-africa-hunger-famine-facts.

Wiebe, K., Robinson, S., & Cattaneo, A. (2019). Climate change, agriculture and food security: Impacts and the potential for adaptation and mitigation. In C. Campanhola & S. Pandey (Eds.), *Sustainable food and agriculture: An integrated approach* (pp. 55–74): London: Elsevier.

World Bank. (2010, December). *Cities and climate change: An urgent agenda* (Urban Development Series Knowledge Papers [63704, 10]). Urban Development Series Knowledge Papers, Washington, DC.

Yer, H. (2017). The effects of successful ICT-based smart city services: From citizen's perspectives. *Governemnent Information Quarterly, 34*(3), 556–565.

Ylä-Anttila, T., Gronow, A., Stoddart, M. C. J., Broadbent, J., Schneider, V., & Tindalle, D. (2018). Climate change policy networks: Why and how to compare them across countries. *Energy Research & Social Science, 45,* 258–265.

Zhang, Y.-X., Chao, Q.-C., Zheng, Q.-H., & Huang, L. (2017). The withdrawal of the U.S. from the Paris Agreement and its impact on global climate change governance. *Advances in Climate Change Research, 8*(4), 213–219.

Zhao, P., & Zhang, M. (2018). The impact of urbanisation on energy consumption: A 30-year review in China. *Urban Climate, 24*, 940–953.
Zhu, Y. (2017). *The urban transition and beyond: Facing new challenges of the mobility and settlement transitions in Asia*. Paper presented at the United Nations Expert Group Meeting on Sustainable Cities, Human Mobility and International Migration, New York.

CHAPTER 2

Decarbonization and Urban Sustainability

Abstract Why climate change mitigation plans often fail to gain global consensus is due to geo-economic political influences. Where the question often arises of why should countries, that contribute less to climate change, made to suffer the most, and why countries that benefit economically the most from climate change, are now asking others to reduce economic activities on the basis of reducing pollution emissions. This chapter introduces the concept of regional decarbonization. It outlines how several attempts in proposing climate change mitigation plans have attracted resistance from developed economies. These negative responses have occurred even though proposed alternative models can theoretically work and often call for a systemic shift that can disrupt global economies prompting a perception that they are a threat that governments are not always apt to accommodate.

Keywords Climate change · Regional decarbonization · Climate · Policy · Sustainability · Cities · Resilience

Z. Allam et al., *Cities and Climate Change*,
Palgrave Studies in Climate Resilient Societies,
https://doi.org/10.1007/978-3-030-40727-8_2

Introduction

As the role and impact of cities upon the global economy get more pronounced, the discourse on climate change is also increasingly gaining traction. To start with, cities are increasingly being recognized as the greatest recipients of negative impacts of climate change. Van der Heijden, Patterson, Juhola, and Wolfram (2018) affirm this by highlighting how different cities are experiencing water shortage and quality challenges resulting in reduced supplies of basic resources like foodstuffs, medical supplies, potable water and shelter. The latter is prompting increases in water access and quality costs. Cities are also experiencing issues like increased incidences of flooding, droughts and the emergence and spread of tropical and no-tropical diseases that individually and or collectively have far-reaching negative impacts upon urban infrastructure, water supply and management, energy distribution, housing and liveability status of urban areas, and thus human, animal and vegetation well-being. On this topic, although some areas and cities have reputedly little or no contribution to this global menace, Mi et al. (2019) showcase that a sizeable number of cities are on the front line when it comes to contributing to climate change. The UN Environment (2019) highlights that cities overall are estimated to contribute over 75% of the total CO_2 emissions, especially following their massive consumption of items like energy, construction materials and other natural resources. Marsden and Rye (2010) also attribute this to the swelling demand for transportation and manufacturing sectors where machineries and automobiles rely heavily upon non-renewable products. Such trends in urban areas have seen increased calls for climate action at both local, national, regional and international levels (United Nations, 2017a). Due to this, there are notable international accords that have been cemented with the latest one being the *Paris Agreement* (UNFCCC, 2018) that seek to meditate and redress these trends.

However, even with the formulation of progressive climate policies and *Agreements*, their implementation is sometimes ineffective because some governments perceive such strategies being a means to disrupt or derail their 'natural' economic growth processes and thus, if faithfully implemented, would dramatically negatively impact upon their economic prosperity. On this, Meng and Rode (2019) demonstrate that there are numerous litigations that have been filed concerning climate change, with over two-thirds accusing governments, especially those in developed economies, of neglecting their Agreement signatory and or legislative obligations and responsibilities in this realm. According to several authors, most of these governments do not 'on the ground' adhere to local or national climate mitigation policies, but operate on the hope and belief that their counterparts in other countries would implement such measures faithfully. Therefore, locally, processes and programmes are pursued even those with high-emitting capacity, that meet local economic and policy goals despite their international stage rhetoric. When pressured by the international community for environmental conservation actions, such governments threaten to withdraw from their commitment on internationally agreed climate mitigation agendas as demonstrated by the recent actions of the US with the *Paris Agreement* (Y.-X. Zhang, Chao, Zheng, & Huanga, 2017).

This lack of enforcement of internationally agreed climate change policies leads to an unfair economic landscape where rich countries (and their cities within) keep getting richer while less developed ones (and their cities within) continue to suffer the consequences of climate change mostly instigated by their rich counterparts. This truth is shared by Diffenbaugh and Burke (2019) who argue that the GDP per capita of poor countries would probably be higher were it not for the evident impacts of climate change. They argue that the economic landscape is disproportionally skewed in favour of rich economies that historically have had the advantages and latitude to benefit directly from their exploitation of fossil fuels and cheap colonial resources to develop their own economies (Tharoor, 2016). Colonized countries have been particularly susceptible to this resource exploitation to their own economic advancement or re-advancement.

Indeed, on the issue of economic status, there has been a noticeable relationship between GDP and emissions (Cohen, Jalles, Loungani, Marto, & Wang, 2019; Magazzino, 2016) as shown in Fig. 2.1.

Unsurprising, Fig. 2.1 highlights that higher-emitting countries have achieved economic prosperity while riding on unsustainable and exploitative practices, furthering a model which is pursued by many others; including less developed countries and those facing the impacts of climate change. From this pursuit, Aye and Edoja (2017) highlight that the noticeable economic growth of various countries has had negative implications on greenhouse gas (GHG) emissions, especially due to their reliance upon unsustainable practices like energy production from fossil fuels and other environmentally unfriendly sources, and their wanton extraction of resources from the environment. Such practices characterize the development model of most developed economies, and developing economies are also seen to be following in the same footsteps (Johnsson, Kjärstad, & Rootzén, 2018). A majority of these countries,

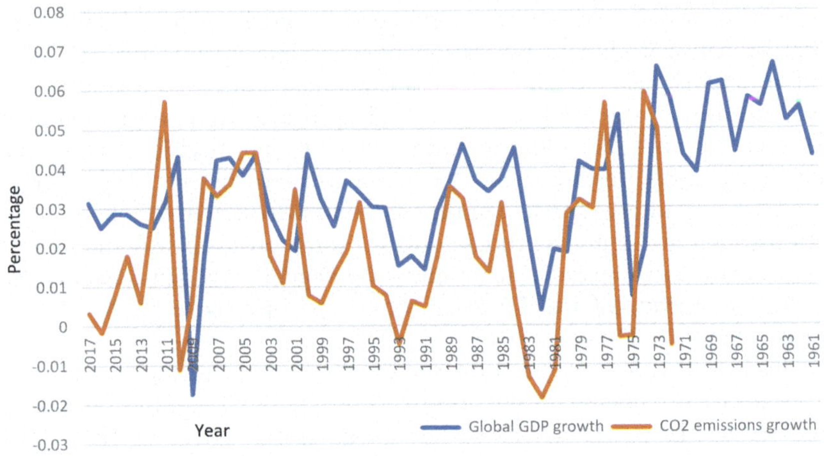

Fig. 2.1 Relationship between GDP and CO_2 emissions (*Data Source* UN, 2019)

as reported by Covert, Greenstone, and Knittel (2016) are also increasing their consumption of fossil fuels, expanding their manufacturing sectors, through the same reliance upon unsustainable practices. The authors argue that such practices are a consequence of the dearth of viable and evidence-supported policies formulated that aim at limiting the use of non-renewable fuel consumption. As a result, the global consumption of fossil fuels has been recorded to have reached over 7.5%. Unfortunately, these less developed countries, as highlighted above, are the ones most highly impacted from climate change effects, and because of their adoption and continued practice of unsustainable frameworks climate change resilience will be far from being achieved.

The above background points to the fact that there needs to be a better understanding of this context where an economic geographical and political imbalance has been created through the transmissions of emissions. These transmissions result in an emerging issue of climate policy being influenced and driven by countries that have been reported to be the highest polluters.

Meng and Rode (2019) evidence this in their examination of trends in the enactment of domestic climate change policies in the US, one of the largest emitters of this century. For example, the *American Clean Energy and Security Act* bill, proposed in 2009, that is often highlighted as being the most promising pieces of legislation tabled in the US Congress in combating local emissions. However, many US local high-polluting firms invested over USD$700 million, as reported by McMahon (2019), to lobby against the bill, that eventually failed. Such political lobbing is common on varying matters across the globe, where powerful anti-climate sentiments from powerful economies and or corporations, like the example of the US withdrawing from the *Paris Agreement*, have crowded global climate policies.

On this, even when strategies like carbon taxation have been introduced, like the popular use of the Emission Trading Scheme (ETS) in Europe, it has been observed that adoption has been influenced greatly by powerful economies and or corporations, and that Europe's geopolitical landscape of climate change initiatives has not managed to bring equity in climate change responses and outcomes. In reality, with such taxation systems or mechanisms, higher emitters are supposed to contribute more to mitigation funds. But, as noted by Haite et al. (2018), this has not been the case. Least developed and developing economies have been seen to struggle to secure sufficient funds to invest in mitigation infrastructures. Yet, such would be more available if polluting economies were honouring their obligations to compensate for their continued emissions. Therefore, as profiled in the next section, such trends need to be examined on the larger global sphere before tailoring localized solutions.

Regional Decarbonization

The imbalance in emissions generation between regions is increasingly apparent, and it is now an emerging field of research. Chancel and Piketty (2015) highlight that it is important to understand the underlying issues that promote such a imbalance, as this will help to strategize on how to handle each region so as to achieve the targets set in the *Paris Agreement*. On this, various models employed in determining the contribution of emissions in different regions have shown that economies in North America, Oceania, Latin America and Europe contribute more emissions than their counterparts in Asia and Africa as shown in Fig. 2.2.

The above conclusion about regional imbalances is true, whether the emissions are measured in terms of population or in the GDP of different countries. Ritchie (2018), in terms of income, has pointed to the fact that the combined richest and middle-income countries account for over

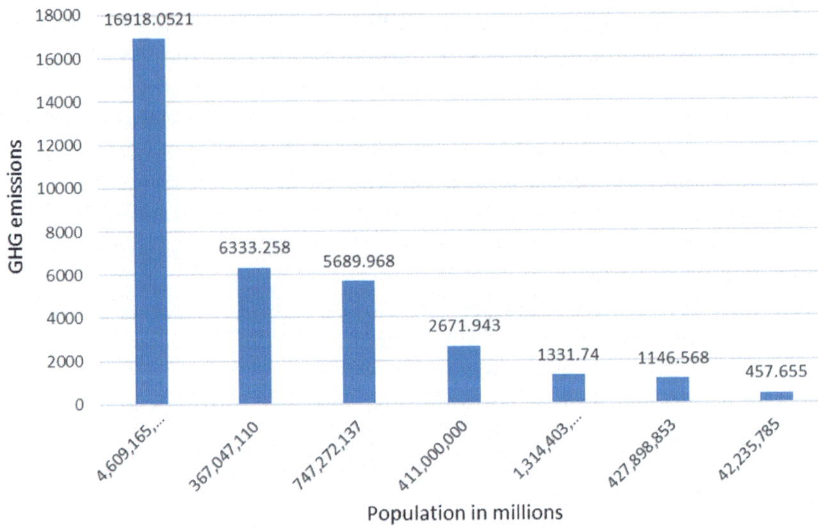

Fig. 2.2 Regional contribution of emissions (*Data Source* Worldometers, 2019)

86% of the global CO_2 emissions, while lower-income economies only account for 14%, and their poorer counterparts are only responsible for 0.5% of the global economy. In terms of population, people in higher-income economies contribute more emissions than those in the middle- and low-income economies. For example, it is reported that Asian countries, having an aggregated population of more than 60% of global population, only account for approximately 49% of the global emissions. In contrast, North America, home to only 5% of the global population, is responsible for more than 18% of the total global emissions (IEA, 2019). Africa, on the other end, has the lowest emissions of only 4% despite being home to over 16% of the world population (Ritchie, 2018).

Looking at the urgency on climate change, and the fact that the world is grappling with rapid urbanization and unsustainable concentrations of people in existing and emerging urban areas, the emission quotas of certain regions are expected to rise creating further global imbalances. To put this in perspective, it has been observed that the world's megacities (those with more than 10 million people) are increasing in numbers and are projected to climb from the current 31 cities to some 43 cities by 2050. The same trend will be even more pronounced in regard to big cities (those with a population of more than 1 million people, but less than 10 million) that are projected to increase from the current 520 cities to over 660 cities by 2050 (United Nations, 2017b). The increase in such numbers will mean an exponential increase in the consumption of diverse resources in the process of generating energy that is a major contributor of emissions.

Looking at the various emission sources, the energy sector, according to a IRENA, IEA, and REN21 (2018) report, accounts for over 70% of all the emissions. An increase in demand in this production would continue to exacerbate these challenges, especially if the call for economies to shift to renewable energy sources is not urgently heeded. In addition to the energy sector, urbanization and population booms in cities will continue to exert pressure upon urban manufacturing sector, and this is seen as the second highest emitting sector ahead of the transportation sector that parallels population growth. Abdoli, Rezaee, and Hasanian (2016) further argue that the concentration of humans in urban areas has increased incidences of emissions from the increased wastes generated from households, industries and offices especially in megacities that are located in developing economies. Abdoli et al. (2016) attribute this to the expansive industrializations taking place in these regions driven by insatiable demands for manufactured goods by populations whose incomes continue to grow. Increases in these activities, when compounded, mean that the emissions from these regions will continue to increase relative to the least developed economies. Thus, we appear to be continuing to fuel the global emission generation imbalance that already exists by enabling major resource exploitations to generate energy to service these expansive industrializations.

On the above, it is worth highlighting that while the said imbalance depicts an unsustainable social landscape, it does not mean that least developed economies should be allowed to continue these activities unabated while their counterparts in developed and developing economies impose strong deterrents and policies. This notion is well captured in the *Paris Agreement* where the responsibility to reduce emissions is shouldered on all economies regardless of their development status. On this, Sahu and Saizen (2019) highlight that it would be inaccurate to overlook the potential of developing countries in contributing to emissions generation, and giving them a wide berth while closely monitoring the actions of developed countries would be missing the mark and our responsibility in enabling the fulfilment of the *Paris Agreement*.

The authors argue that the determination of potential to emit should not be based on past emissions only, but should also consider the population and development potential of all. By doing this, the determinations would allow climate mitigation actors to treat least developed economies on an equal footing. This viewpoint is shared by Pan, Teng, Tian and Wang (2015) who highlight that such actors can take advantage of the wide range of countries' emission sharing frameworks to ensure that equity is served even in the case of developing economies. Taking this option would ensure that least developed economies are not stifled on their much-needed development agenda, while at the same time, they are protected from impacts arising from the actions from developed and developing economies. The equity principle is to be applied to ensure that developed economies and developing ones adopt ambitious reduction targets commensurate to their emission potential, while least developing ones are also made to make their contribution no matter how meagre those are perceived (Pan, den Elzen, Höhne, Teng, & Wang, 2017). According to Ari and Sari (2017), the need for collective participation is also to be viewed from the reality that there are challenges in differentiating developing, developed and least developed economies, and giving some the leeway to abstain participation would erroneously allow high emitters to continue unabated. For this reason, the authors support the need to have a clear definition and criteria depicting where every country lies in regard to climate responsibility, so that policies can be aligned appropriately.

Climate Policy and Decarbonization

The process of decarbonization is not a straightforward affair and requires shrewd strategies. But, it would be worth starting by understanding the contributions of specific industries. Thus, act from a position of information. This is especially necessary as each industry contributes differently to climate change and a generalized discourse on the need to decarbonize would be defeated. But, when the contribution to climate change for each sector is computed independently, it could allow the actors to know where maximum attention needs to be focused. This is true, especially considering that the energy sector has for long been associated with being an emission-intensive industry, with the construction industry and the transport sector following in its heels. Gerres, Chaves-Ávila, Linares, and Gomez (2018) highlight that the process of decarbonization has different options that can be adopted, but come as sector-specific options due to the different requirements that need to be implemented. Such requirements include relevant infrastructures and technologies that are, again, sector specific. This means that the right decarbonization strategy needs to be applied at the earliest opportunity so as to reduce inefficiencies and unnecessary costs that may be experienced at a later date if the process was to be undertaken without consideration of all varying sectors. Tilburg et al. (2019) support that the clarity on sectoral contributions and need for decarbonization is essential to guide domestic implementation strategies since different economies and regions have specific sectors and industries that demand varying decarbonization attention. Similarly, the issues of financing that could be better managed as budget provisions are usually made according to sector sensitivity.

Emissions from different industrial sectors arise in different forms and intensities. The energy sector is understood to be the leader due to its over-reliance on fossil fuels for the production and transmission of energy and also due to its demand on other sectors of any given economy. Zhou, Wang, Yuan, and Ou (2018) highlight that industrial sectors account for more than half of the energy demand, especially in developing economies, and that the decarbonization processes in these sectors can only be achieved if a shift from non-renewable sources is adopted in addition to the embracement advanced technology in this sector (Bataille et al., 2018).

With respect to the construction industry, its contribution to emissions is based upon its heavy reliance of the energy sector in almost all phases of any given project. For the construction industry, even in the extraction and manufacturing of raw materials, energy is still very important, thus making this sector a potential candidate for decarbonization. Conci, Konstantinou, van den Dobbelsteen, and Schneider (2019) argue that the road to decarbonization in this sector can be achieved by emphasizing innovative building strategies, and using power generated from renewable sources. This argument can also be seen in the transportation sector, as narrated by Ding, Jin, Li, and Wang (2013), where a sectoral analysis shows that most emissions arise from the use of fossil fuels to power automobiles and similar engines and also derive from the reliance on the energy sector especially during infrastructural development. In the agricultural sector, Tubiello (2019) concludes that emissions from both crop and livestock activities are significant and need to be addressed the same way to that in the energy and transportation sectors.

Decarbonization efforts in the above sectors can be addressed by adopting specific decarbonization policies that have been tried and tested in a number of regions. The most prominent of these policies is the carbon pricing. Bassi, Carvalho, Doba, and Fankhauser (2017) argue that carbon pricing can be very effective as it targets specific carbon emitters directly. Thus, it seeks to deliver targeted emission reduction through economic mechanisms. Technology-specific subsidies are another policy initiative that promotes the use of advanced emission reduction technologies in different sectors. Those user who are entitled to receive these subsidies need to heed calls to adopt technologies that promote the reduction of emissions, and these adoptions have been seen to increase competitiveness and the adoption of sustainable practices, especially in the energy sector, with renewable sources being exploited more including in oil-rich regions (Purkus, Gawel, & Thrän, 2017).

The Global Ecosystem

There are numerous models that cater for regional decarbonization and these have been fabricated respecting different regional factors that can be comprehended by critically looking at how emission accounting in different regions is performed, and how those calculations are reflected on the global sphere. On this, Karakaya, Yilmaz, and Alataş (2018) posit that in most cases, emission accounting is normally based upon the Production-Based Accounting (PBA) system. But PBA system has been the recipient of criticism for its failure to account for emissions arising from trade. To explain, the criticism is that the accounting system only has regard for emissions that arise from the domestic production of goods, whether in farms or industries, inside a given country or region, and does not consider emissions from energy-intensive products acquired and transmitted from another country (or region), and consumed in the country where the accounting is taking place. Such emissions cannot be wished away as it is estimated that such accounts for between 25 and 30% of the global emissions (Z. Zhang, Zhu, & Hewings, 2017). For this reason, it is said that the PBA system is not holistic. Thus, emission accountants need to incorporate the Consumption-Based Accounting (CBA) system that accounts for any form of emissions in a country, including those that are emitted by goods imported into the country as noted by

Peters, Andrew, and Karstensen (2016). Models formulated on the basis of PBA system then cannot be assumed to have the potential to achieve the intended decarbonization targets. On this line of argument, Dobson and Fellows (2017) have evidenced that it is advantageous for countries, whose emission-intensive products are consumed elsewhere, to use the CBA system to continue producing them unabated. Nevertheless, for the purposes of equity and fairness, these models need to be formulated based upon the CBA, such that decarbonization targets reflect environmental realities.

To remedy shortcomings in the PBA model, there are numerous ecosystem accounting models that set out to map not only emission generations, but also map carbon captures by various ecological assets around the world. These ecosystem accounting models include spatially explicit models that map ecosystem service flows to economies and how such flows impact societies. These ecosystem accounting models are dependent upon mapping all the components of the ecosystem thus making it possible to identify the drivers that affect the actual flow of services (Vallecillo, La Notte, Zulian, Ferrini, & Maes, 2019). Another ecosystem accounting system is the operational model advanced by Cowling et al. (2008) that is focused upon mainstreaming ecosystem services development and implementation through identifying social, biophysical and valuation opportunities and constraints that normally crowd ecosystem assets. Other ecosystem accounting systems include the Ecosystem Services Valuation (ESV) accounting model advanced by Q. Yang et al. (2018) that focuses upon, amongst other things, bridging the non-monetary and economic values of ecosystem assets and assessing the return of wealth from nature. By doing this, the ESV model supports that extra caution on emissions should be given priority to safeguard these ecosystems as they are being equated as economic assets. This caution can be achieved by employing carbon capture and storage models like the revenue models that help in earning income from captured and stored carbon enabling these revenues to be used in mitigation programmes targeting the safeguarding of ecological assets. Kheshgi et al. (2009) argue that having a sound ecosystem accounting model is a prerequisite for having a viable carbon capture and storage model that would not only promote mitigation, but also improve the revenue earned from mitigation programmes. The availability of the above models provides us with a global picture of how the world works, where emissions are being emitted, and where and how much is being absorbed back.

While policies are being devised and implemented to encourage decarbonization, it is argued that more policies need to be drafted in order to protect global assets that capture carbon. This is because these global assets that capture carbon play a critical role in the survival of the global ecosystem. Hurteau, North, Koch, and Hungate (2019) posit that such policies are required urgently since most global assets, especially those in the forests and aquatic-terrestrial environments are constantly being disturbed and destabilized by events including droughts, global warming, wildfires, flooding and human-induced impacts such as logging and intensified changes in land uses. Policies should then focus upon promoting afforestation, reducing deforestation and forest degradation, and increasing forest carbon density. Goldstein et al. (2012) advance this narrative by arguing that new policies ought to be hinged on the need for a trade-off between private and public interests when dealing with ecosystem assets, and by doing this, the trade-offs manage to actualize goals such as climate change mitigation, economic and societal benefit diversification as well as attaining food security. On this point, it is argued that the public interest, those that touch on protecting global assets like food should supersede private interests that are majorly underpinned by selfish profit-making interests, more so by politicians and corporations. This is demonstrated by Johnson (2019) who argues that geographical areas, such as coastal zones that have substantial capacity to contain unprecedented amounts of carbon, should be prudently managed by integrating vegetation (such as mangroves, seagrasses and other vegetation) instead of destroying the existing vegetation communities for selfish private gains in the form of luxury resorts and coastal housing estates. On this, policies that target coastal wetland protection should be even more stringent, since, besides allowing for carbon capture they promote coastal storm protection and improve water quality. More importantly, they have the potential to accommodate vegetation, that is declining inland as more land conservation reserves are being converted into productive agricultural lands or built environment landscapes.

The formulation of new policies has never been an issue, as there are numerous tools and information available to assist on this front. Nevertheless, there are always some challenges that linger when such policies are proposed. In particular, the issue of economics. That is, how to achieve consensus on climate efforts so that fund-raising can be strategically made available towards the protection of global ecosystem assets is one area that needs to be overcome. This challenge, as Gamarra et al. (2019) denote, has led to undervaluing ecological assets. Thus, little financial returns are provided to safeguard them, while the lion share from revenue earned as a result of carbon capture and other capital flows from them, is utilized in private sectors. Therefore, as the aforementioned authors support, policies formulated for the purposes of protecting these assets need to vividly make it clear how the returns and values generated from these assets are proportionally shared for private and public interests.

Paving Way for the City

While the role of protecting ecological assets, as described in the previous section, is apparent and inevitable, these ecological assets face numerous challenges from various quarters, with urbanization and demographic boom being the most prominent ones (Z. Allam, 2020a, 2020b, 2020c; Parris et al., 2018). Urban population increase has been seen to exert pressure upon existing cities especially in regard to settlement areas and infrastructural development. For this reason, it very evident that city boundaries are expanding beyond the set perimeters and into ('green belt', agricultural, passive conservation and recreation, etc.) reserves areas due to urban sprawl. These unsustainable trends have opened a plethora of challenges, with the loss of biodiversity life at alarming rates, thereby threatening landscape preservation and ecological conservation agendas. It is also visually evident that ecological assets, like potable and saline water resources, are facing wanton compromise. Such negative trends affect hydrological cycles and the quality of water that both humans, animals and plants heavily rely upon (Z. Allam, 2012, 2017; Z. Allam & Jones, 2018; McGrane, 2016). Unfortunately, there is no visible end to these problems, especially if there is no drastic 'development' paradigm shift, that could result in mainstream or Tradition Indigenous, or a mix of both, sustainable practices being urgently embraced.

This conclusion is true when one reviews future projections by the United Nations (2017b) that conclude urban population concentrations and agglomerations will continue to increase to over 68% by 2050. This conclusion is borne out of the realization that it is difficult and outright near impossible to contain rural-to-urban migrations and this projected urban population boom thereby pointing to the validity of pursuing practical regional decarbonization policies that address and run in parallel to this urbanization challenge.

Indeed, despite the gloomy and uncertain future on this topic, there have been some flashes of hope, as explained by Gomez Echeverri (2018), that emanates from the optimistic adoption of advanced technological practices like smart cities, and the slow but sure shift in favour of renewable energies in a number of economies. The two mentioned examples, plus the increased call from different quarters demanding urgent action on the issue of climate change, if embraced conventionally can bring some notable progress on the formulation of decarbonization policies.

However, these policies are not always welcomed and have been seen to have sparked some conundrums especially with respect to the ethical and moral grounds of whether biodiversity conservation should supersede the welfare of urban life. On this, Hampicke (1994) posits that it is the ethical and moral obligation of the current generation to conserve biodiversity for the sake of future generations, even though the current generation is already overburdened with its ethical responsibility of combating the impacts of climate change by the actions and inactions of its generation. These views are shared by Pearson (2016) who argues that discourses on conservation versus human welfare are grounded in human's selfish nature, which crowds human perceptions to the point of apportioning monetary value to that of biodiversity. By so doing, their ignorance belittles the potential and contribution of given ecological assets, but they forget that some of their destructive actions are irreversible and carries priceless negative repercussions upon the liveability status of entire human, animal and vegetation communities. As evidence, Melliger, Braschler, Rusterholz, and Baur (2018) demonstrate that it is possible for the world's population to co-exist with the environment in a symbiotic relationship, and thus quench ethical and moral dilemmas. Luck et al. (2012) advise that such co-existence should be entrenched in the consideration of policies, planning and formulation whereby ecological assets conservation and protection are not be viewed as an afterthought, but a deliberate and conscious action. This strategy would help alienate the bias of viewing biodiversity as being inferior to human welfare and enculturate the idea that biodiversity is an important facet of liveability especially in urban settings (Chan, Satterfield, & Goldstein, 2012; Tammi, Mustajärvi, & Rasinmäki, 2017).

To conclusively ensure that both urban life and biodiversity life receive respectable and equitable attentions, and the ethical and moral dilemma that surrounds their relationships are holistically addressed, global metabolism also needs to be studied. The immediate conclusion is that a global and more inclusive emissions accounting system is paramount in its construction because this would allow the factoring of both production and consumption emissions together with measurements of emissions capture in the context of cities. Such a model and formula will provide us with a fresh overview on how to more deeply examine and measure decarbonization from the outset of city development and growth in the global context. In the long run, this formula and model would prove very important, since, as discussed in the previous section, and echoed by Chinowsky et al. (2011), some cities in developed countries are eating large amounts of resources, while others—especially those in developing and least developed economies—are experiencing climate-related impacts from actions of others. Therefore, a global outlook on decarbonization should encompass biodiversity in the context of cities, and be all-encompassing regardless of city size, ecosystem types, geographical locations, and the city's direct or indirect interactions with global ecological assets.

Climate Change Mitigation and Regional Decarbonization

While decarbonization in the context of cities is important in ensuring that a sustainable transition is achieved in the long and medium terms, it is of critical importance to address the pressing challenges of the immediate short term. This includes especially those challenges instigated by the impacts of climate change that are being increasingly felt in urban areas. This is paramount because vulnerable regions, more so those in least developed economies, have been found to have wanting capacity to withstand such impacts that are manifest in various ways including loss of life, destruction of properties, human and animal displacements, and the reduction of basic supplies like food, clean potable water and medical supplies. For this reason, the pursuit of urban resilience is seen as key in the survival of urban fabrics, both in the short and long terms, as this would help vulnerable regions to withstand the aforementioned immediate impacts while decarbonization agendas are being pursued. Dodman, Archer, and Satterthwaite (2019) highlight that there is a need for short-term planning for climate eventualities as it allows stakeholders to make immediate financial provisions for resilience programmes. Similar arguments are made by Meyer and Schwarze (2019) who argue that failing to invest in infrastructures that strengthen and boost urban resilience only increases the vulnerability and the cost of reconstruction actions post-climate change incident. Indeed, from this basis, it can be argued that decarbonization cannot be effectively achieved if the short-term impacts of climate change are not mitigated and addressed appropriately. Yang, Dietz, Yang, Zhang, and Liu (2018) agree on this arguing that when short-term impacts are overlooked, they eventually turn to mid-term and long-term challenges that become daunting to address, both in terms of cost and capacity.

The risks of failing to invest in urban resilience manifest in numerous ways, with the financial burden being amongst the pressing challenges that most cities face (M. Z. Allam, 2019). From a wide scoping of literature (Attapattu & Padmasiri, 2018; Daseking & Kozack, 2003; Estache, Serebrisky, & Wren-Lewis, 2015; Payer, 1975), it is clear that the financing of climate change mitigation is a question that is increasingly being explored, as some of the existing financing models have been argued to lead to unsustainable debt cycles. In some developing and least developed economies, the problem of debt has been seen to persist, with some of these economies gambling with some of their critical public infrastructures to service lenders on failure of repayments or as economic securities. In other cases, as reported by Weerakoon (2017), critical public service deliveries have stalled or have been critically affected by lack of sufficient finances because a large portion of available financial resources is diverted towards servicing loans instead of providing for urban dwellers. In addition, as demonstrated by the Brettonwoods Project (2019), these countries are also increasingly finding themselves trapped in debt obligations to service loans due to unforeseen external climate change incidents. This economic cycle is very evident in the case of Mozambique that is now in heavy debt following the devastating Cyclone Adai in March 2019.

On this, although the *Paris Agreement* supports that recoveries from such disasters be funded from financial advances from international communities in form of aid, this has failed to materialize in Mozambique and the potent original source of finance were loans advanced by the International Monetary Fund (IMF) (Ehrenreich, 2019; Sauer, 2019). This example amplifies the friction that exists in climate change mitigation and adaptation financing, as those most responsible, are less affected by the impacts of climate change leading to the practice of being a passive observer as their poorer counterparts wallow in debt arising from having to address climate change incident recovery programmes. In an ideal situation, those high-polluting economies should be the ones compensating for the damages prompted by climate change incidents even in other jurisdictions by their actions. But, unfortunately, as noted by Knight (2011), the amount they contribute to compensate for their actions cannot suffice in financing mitigation and adaptation programmes in vulnerable regions.

On the above, Chapter 3 showcases how economic incentivization mechanisms can be used as a means for capital raising towards climate change mitigation. However, the success of these mechanisms demands upon the cooperation of both public and private sectors which is paramount if we want to be able to both finance climate change mitigation programmes or to implement decarbonization policies.

References

Abdoli, M., Rezaee, M., & Hasanian, H. (2016). Integrated solid waste management in megacities. *Global Journal of Environmental Science and Management, 2*(3), 289–298.

Allam, M. Z. (2019). *Urban resilience and economic equity in an era of global climate crisis.* University of Sydney.

Allam, Z. (2012). Sustainable architecture: Utopia or feasible reality? *Journal of Biourbanism, 2,* 47–61.

Allam, Z. (2017). *A theoretical application of the extended metabolism model in Port Louis in a bid to promote urban sustainability.* Paper presented at the 2nd International Conference on Energy, Environment and Climate Change (ICEECC 2017), Mauritius.

Allam, Z. (2020a). Religion and urban planning. In Z. Allam (Ed.), *Theology and urban sustainability* (pp. 1–19). Cham: Springer.

Allam, Z. (2020b). Religious matrimony, urban sprawl and urban morphology. In Z. Allam (Ed.), *Theology and urban sustainability* (pp. 21–35). Cham: Springer.

Allam, Z. (2020c). Urban and graveyard sprawl: The unsustainability of death. In Z. Allam (Ed.), *Theology and urban sustainability* (pp. 37–52). Cham: Springer.

Allam, Z., & Jones, D. (2018). *Promoting resilience, liveability and sustainability through landscape architectural design: A conceptual framework for Port Louis, Mauritius; A small island developing state.* Paper presented at the IFLA World Congress Singapore 2018, Singapore.

Ari, I., & Sari, R. (2017). Differentiation of developed and developing countries for the Paris Agreement. *Energy Strategy Reviews, 18,* 175–182.

Attapattu, A. M. C. P., & Padmasiri, H. M. N. (2018). Long run effect of public debt on economic growth in Sri Lanka. *Modern Economy, 9*(4), 775–789.

Aye, G. C., & Edoja, P. E. (2017). Effect of economic growth on CO_2 emission in developing countries: Evidence from a dynamic panel threshold model. *Cogent Economics & Finance, 5*(1), 1–22.

Bassi, S., Carvalho, M., Doba, B., & Fankhauser, S. (2017). *Credible, effective and Publicly acceptable policies to decarbonise the European Union.* Retrieved from http://www.lse.ac.uk/GranthamInstitute/publication/credible-effective-publicly-acceptable-policies-decarbonise-european-union-final-report/.

Bataille, C., Åhman, M., Neuhoff, K., Nilsson, L. J., Fischedick, M., Lechtenböhmer, S., et al. (2018). A review of technology and policy deep decarbonization pathway options for making energy-intensive industry production consistent with the Paris Agreement. *Journal of Cleaner Production, 187,* 960–973.

Brettonwoods Project. (2019, July 30). *IMF and World Bank complicit in 'climate debt trap' following Mozambique cyclones.* Retrieved from https://www.brettonwoodsproject.org/2019/07/imf-and-world-bank-complicit-in-climate-debt-trap-following-mozambique-cyclones/.

Chan, K. M. A., Satterfield, T., & Goldstein, J. (2012). Rethinking ecosystem services to better address and navigate cultural values. *Ecological Economics, 74,* 8–18.

Chancel, L., & Piketty, T. (2015). *Carbon and inequality: From Kyoto to Paris—Trends in the global inequality of carbon emissions (1998–2013) & prospects for an equitable adaptation fund.* Retrieved from Paris.

Chinowsky, P., Hayles, C., Schweikert, A., Strzepek, N., Strzepek, K., & Schlosser, C. A. (2011). Climate change: Comparative impact on developing and developed countries. *Engineering Project Organization Journal, 1*(1), 67–80.

Cohen, G., Jalles, J. T., Loungani, P., Marto, R., & Wang, G. (2019). Decoupling of emissions and GDP: Evidence from aggregate and provincial Chinese data. *Energy Economics, 77,* 105–118.

Conci, M., Konstantinou, T., van den Dobbelsteen, A., & Schneider, J. (2019). Trade-off between the economic and environmental impact of different decarbonisation strategies for residential buildings. *Building and Environment, 155,* 137–144.

Covert, T., Greenstone, M., & Knittel, C. R. (2016). Will we ever stop using fossil fuels? *The Journal of Economic Perspectives, 30*(1), 117–137.

Cowling, R. M., Egoh, B., Knight, A. T., O'Farrell, P. J., Reyers, B., Rouget, M., … Wilhelm-Rechman, A. (2008). An operational model for mainstreaming ecosystem services for implementation. *PNAS, 105*(28), 9483–9488.

Daseking, C., & Kozack, J. (2003). Avoiding another debt trap: Low-income countries need a sustainable borrowing strategy if they are to achieve development goals. *Finance & Development*, 20–23.

Diffenbaugh, N. S., & Burke, M. (2019). Global warming has increased global economic inequality. *PNAS, 116*(20), 9808–9813.

Ding, U., Jin, F., Li, Y., & Wang, J. (2013). Analysis of transportation carbon emissions and its potential for reduction in China. *Chisese Journal of Population Resources and Environment, 11*(1), 17–25.

Dobson, S., & Fellows, G. K. (2017). Big and little feet: A comparison of provincial level consumption and production-based emissions footprints. *The School of Public Policy Publications, 10*(23), 1–45.

Dodman, D., Archer, D., & Satterthwaite, D. (2019). Editorial: Responding to climate change in contexts of urban poverty and informality. *Environment and Urbanization, 31*(1), 3–12.

Ehrenreich, B. (2019). *It's time to pay our climate debt to countries like Mozambique*. Retrieved from https://www.thenation.com/article/mozambique-idai-climate-debt/.

Estache, A., Serebrisky, T., & Wren-Lewis, L. (2015). Financing infrastructure in developing countries. *Oxford Review of Economic Policy, 31*(3–4), 279–304.

Gamarra, N. C., Correia, R. A., Bragagnolo, C., Campos-Silva, J. V., Jepson, P. R., Ladle, R. J., et al. (2019). Are protected areas undervalued? An asset-based analysis of Brazilian protected area management plans. *Journal of Environmental Management, 249*, 109347.

Gerres, T., Chaves-Ávila, J., Linares, P., & Gomez, T. (2018). *A review of cross-sector decarbonisation potentials in the European energy intensive industry*.

Goldstein, J. H., Caldarone, G., Duarte, T. K., Ennaanay, D., Hannahs, N., Mendoza, G., et al. (2012). Integrating ecosystem-service tradeoffs into land-use decisions. *PNAS, 109*(19), 7565–7570.

Gomez Echeverri, L. (2018). Investing for rapid decarbonization in cities. *Current Opinion in Environmental Sustainability, 30*, 42–51.

Haite, E., Maosheng, D., Gallagher, K. S., Mascher, S., Narassimhan, E., Richards, K. R., et al. (2018). Experience with carbon taxes and greenhouse gas emissions trading systems. *Duke Environmental Law & Policy Forum, 29*, 109–182. Retrieved from https://delpf.law.duke.edu/article/experience-with-carbon-taxes-and-greenhouse-gas-emissions-trading-systems-haites-vol29-iss1/.

Hampicke, U. (1994). Ethics and economics of conservation. *Biological Conservation, 67*(3), 219–231.

Hurteau, M. D., North, M. P., Koch, G. W., & Hungate, B. A. (2019). Opinion: Managing for disturbance stabilizes forest carbon. *PNAS, 116*(21), 10193–10195.

IEA. (2019). *CO_2 emissions statistics.* Retrieved from https://www.iea.org/statistics/co2emissions/.

IRENA, IEA, & REN21. (2018). *Renewable energy policies in a time of transition*: IRENA, OECD/IEA and REN21.

Johnson, J. (2019, February 25). Capturing carbon: Can it save us? *Greenhouse Gases.* Retrieved from https://cen.acs.org/environment/greenhouse-gases/Capturing-carbon-save-us/97/i8.

Johnsson, F., Kjärstad, J., & Rootzén, J. (2018). The threat to climate change mitigation posed by the abundance of fossil fuels. *Climate Policy, 19*(2), 258–274.

Karakaya, E., Yilmaz, B., & Alataş, S. (2018). *How production based and consumption based emissions accounting systems change climate policy analysis: The case of CO_2 convergence.* Retrieved from Nazilli, Aydin, Turkey.

Kheshgi, H., Crookshank, S., Cunha, P., Lee, A., Bernstein, L., & Siveter, R. (2009). Carbon capture and storage business models. *Energy Procedia, 1*(1), 4481–4486.

Knight, C. (2011). Climate change and the duties of the disadvantaged: Reply to Caney. *Critical Review of International Social and Political Philosophy, 14*(4), 531–542.

Luck, G. W., Chan, K. M. A., Eser, U., Gómez-Baggethun, E., Matzdorf, B., Norton, B., et al. (2012). Ethical considerations in on-ground applications of the ecosystem services concept. *BioScience, 62*(12), 1020–1029.

Magazzino, C. (2016). The relationship between real GDP, CO_2 emissions, and energy use in the GCC countries: A time series approach. *Cogent Economics & Finance, 4*(1), 1152729.

Marsden, G., & Rye, T. (2010). The governance of transport and climate change. *Journal of Transport Geography, 18*, 669–678.

McGrane, S. J. (2016). Impacts of urbanisation on hydrological and water quality dynamics, and urban water management: A review. *Hydrological Sciences Journal, 61*(13), 2295–2311.

McMahon, J. (2019). *The lobbying against one bill cost all of us $60 billion in Climate damages, economists say.* Retrieved from https://www.forbes.com/sites/jeffmcmahon/2019/06/02/lobbying-against-one-bill-cost-all-of-us-60-billion-in-climate-costs-economists-say/#22e6e90b658f.

Melliger, R. L., Braschler, B., Rusterholz, H.-P., & Baur, B. (2018). Diverse effects of degree of urbanisation and forest size on species richness and functional diversity of plants, and ground surface-active ants and spiders. *PLoS ONE, 13*(6), e0199245.

Meng, K. C., & Rode, A. (2019). The social cost of lobbying over climate policy. *Nature Climate Change, 9*(6), 472–476.

Meyer, P. B., & Schwarze, R. (2019). Financing climate-resilient infrastructure: Determining risk, reward, and return on investment. *Frontiers of Engineering Management, 6*(1), 117–127.

Mi, Z., Guan, D., Liu, Z., Viguie, V., Fromer, N., & Wang, Y. (2019). Cities: The core of climate change mitigation. *Journal of Cleaner Production, 207,* 582–589.

Pan, X., den Elzen, M., Höhne, N., Teng, F., & Wang, L. (2017). Exploring fair and ambitious mitigation contributions under the Paris Agreement goals. *Environmental Science & Policy, 74,* 49–56.

Pan, X., Teng, F., Tian, Y., & Wang, G. (2015). Countries' emission allowances towards the low-carbon world: A consistent study. *Applied Energy, 155,* 218–228.

Parris, K. M., Amati, M., Bekessy, S. A., Dagenais, D., Fryd, O., Hahs, A. K., et al. (2018). The seven lamps of planning for biodiversity in the city. *Cities, 83,* 44–53.

Payer, C. (1975). *The debt trap: The international monetary fund and the third world.* New York, NY: NYU Press.

Pearson, R. G. (2016). Reasons to conserve nature. *Trends in Ecology & Evolution, 31*(5), 366–371.

Peters, G. P., Andrew, R. M., & Karstensen, J. (2016). *Global environmental footprints: A guide to estimating, interpreting and using consumption-based accounts of resource use and environmental impacts.* Retrieved from Copenhagen.

Purkus, A., Gawel, E., & Thrän, D. (2017). Addressing uncertainty in decarbonisation policy mixes—Lessons learned from German and European bioenergy policy. *Energy Research & Social Science, 33,* 82–94.

Ritchie, H. (2018, October 16). *Global inequalities in CO_2 emissions.* Retrieved from https://ourworldindata.org/co2-by-income-region.

Sahu, S., & Saizen, I. (2019). Emissions sharing observations from a diverse range of countries. *Sustainability, 11*(4013), 1–15.

Sauer, N. (2019). *Mozambique 'faces climate debt trap' as Cyclone Kenneth follows Idai.* Retrieved from https://www.climatechangenews.com/2019/04/26/mozambique-faces-climate-debt-trap-cyclone-kenneth-follows-idai/.

Tammi, I., Mustajärvi, K., & Rasinmäki, J. (2017). Integrating spatial valuation of ecosystem services into regional planning and development. *Ecosystem Services, 26,* 329–344.

Tharoor, S. (2016). *Inglorious empire: What the British did to India.* London: Hurst & Co.

Tubiello, F. N. (2019). Greenhouse gas emissions due to agriculture. In P. Ferranti, E. M. Berry, & J. R. Anderson (Eds.), *Encyclopedia of food security and sustainability* (pp. 196–205). Oxford: Elsevier.

UN. (2019). *World economic situation and prospects.* Retrieved from https://www.un.org/development/desa/dpad/wp-content/uploads/sites/45/publication/wesp_mb123.pdf.

UN Environment. (2019). *Cities and climate change*. Retrieved from https://www.unenvironment.org/explore-topics/resource-efficiency/what-we-do/cities/cities-and-climate-change.

UNFCCC. (2018). *COP 24 agenda as adopted*. Retrieved from https://unfccc.int/node/185180.

United Nations. (2017a). *Urban environment related mitigation benefits and co-benefits of policies, practices and actions for enhancing mitigation ambition and options for supporting their implementation*. Retrieved from https://unfccc.int/sites/default/files/resource/docs/2017/tp/02.pdf.

United Nations. (2017b). *World population prospects: Key findings & advance tables*. New York, NY: Department of Economic and Social Affairs, Population Division.

Vallecillo, S., La Notte, A., Zulian, G., Ferrini, S., & Maes, J. (2019). Ecosystem services accounts: Valuing the actual flow of nature-based recreation from ecosystems to people. *Ecological Modelling, 392,* 196–211.

Van der Heijden, J., Patterson, M., Juhola, S., & Wolfram, M. (2018). Special section: Advancing the role of cities in climate governance—Promise, limits, politics. *Journal of Environmental Planning and Management, 62*(3), 365–373.

Van Tilburg, X., Lamboo, S., Luijten, J., Minderhout, S., Emmrich, J., & Roeser, F. (2019). *Sector decarbonisation pathways from different angles: Technology, modelling, politics*. Retrieved from https://ambitiontoaction.net/wp-content/uploads/2019/06/NDC-Update_June2019_web_c.pdf.

Weerakoon, D. (2017). Sri Lanka's debt troubles in the new development finance landscape. *Third World Thematics: A TWQ Journal, 2*(6), 744–761.

Worldometers. (2019). *Population*. Retrieved from https://www.worldometers.info/population/.

Yang, H., Dietz, T., Yang, W., Zhang, J., & Liu, J. (2018). Changes in human well-being and rural livelihoods under natural disasters. *Ecological Economics, 151,* 184–194.

Yang, Q., Liu, G., Casazza, M., Campbell, E. T., Giannetti, B. F., & Brown, M. T. (2018). Development of a new framework for non-monetary accounting on ecosystem services valuation. *Ecosystem Services, 34,* 37–54.

Zhang, Y.-X., Chao, Q.-C., Zheng, Q.-H., & Huanga, L. (2017). The withdrawal of the U.S. from the Paris Agreement and its impact on global climate change governance. *Advances in Climate Change Research, 8*(4), 213–219.

Zhang, Z., Zhu, K., & Hewings, G. J. D. (2017b). A multi-regional input–output analysis of the pollution haven hypothesis from the perspective of global production fragmentation. *Energy Economics, 64,* 13–23.

Zhou, S., Wang, Y., Yuan, Z., & Ou, X. (2018). Peak energy consumption and CO_2 emissions in China's industrial sector. *Energy Strategy Reviews, 20,* 113–123.

CHAPTER 3

Climate Change Mitigation and Urban Liveability

Abstract Countries that pollute mostly are seen as economically more resilient than those that pollute less, and which are also more economically vulnerable and less apt to invest in climate change mitigation programmes. Faced by the impacts of climate change, this lack of financial capacity from less economically resilient countries poses a threat to their urban liveability levels just as much as their urban infrastructures. Due to this, cities turn to international organizations and signed accords to call for global consensus on addressing the climate change phenomenon. This chapter therefore dwells into the various calls from developing countries to address climate change globally, exploring the possible financing pathways and the relationships between climate change mitigation and urban liveability.

Keywords Climate change · Climate change mitigation · Resilience · Urban liveability · Sustainability · Financing

Introduction

Over this twenty-first century, the world has experienced almost unfathomable challenges posed by increasing incidences and calamities arising from the effects of climate change. Specifically, those countries in the

Z. Allam et al., *Cities and Climate Change*,
Palgrave Studies in Climate Resilient Societies,
https://doi.org/10.1007/978-3-030-40727-8_3

Global South have been faced with extreme weather conditions, like heat waves, strong and frequent hurricanes and cyclones, increased and erratic precipitation falls, prolonged and unseasonal winters, prolonged droughts and periods of over 40 °C temperatures, increased incidents of existing and new disease outbreaks and economic hardship generally (IPCC, 1997; Ray et al., 2019; UN Environment, 2019). While these challenges have also been experienced in the Global North, their impacts have been much less in proportion and management capacity considering that these countries are far much better placed in terms of infrastructural development, economic growth and their financial capacity to 'weather' and recuperate from such incidents. In context, the same countries are, unfortunately, those that have been the greatest contributors to climate change as they pursue economic models that, for long, have had little to do with environmental sustainability. As noted by Lanfranchi, Herrero, Palenzuela, Camilloni, and Bauer (2018), economic activities in most developed countries have the capacity and potential to pollute the environment. Such activities are and were historically stimulated by waves of colonial and self-industrializations that allowed these countries to enjoy substantial economic rewards from the numerous industries and manufacturing plants they installed, from the cheap raw material they exploited from pre-developing countries, coupled with the increased global demand for the manufactured products. On this, it is evident that a majority of these industries were, and some still today, relying upon fossil fuels and other non-renewable energy sources to drive their processes. These sources and energies have been accused of emitting massive amounts of greenhouse gases (GHG) that subsequently contribute greatly to climate change (Covert, Greenstone, & Knittel, 2016; Henriques & Sadorsky, 2018; Johnsson, Kjärstad, & Rootzén, 2018). In the current era, consumer behaviours in most developed economies are also to blame for their direct and indirect contributory roles in climate change. The demand for manufactured goods, automobiles, electronics and products sourced from forest reserves, extracted mineral resources and marine ecosystems have been on the rise in those economies, and increasing demands for these resources possess direct correlations to excessive power and fuel consumptions and ecology degradation that are amongst the leading contributors to climate change. These variables have also influenced the reduction of the number of green spaces available inside cities and are also responsible for the massive amounts of waste that are generated and unsustainably dumped each year.

This economic model, which emphasizes the growth of Gross Domestic Product (GDP) with little or no concern on CO_2 levels and other GHG emissions, has long been seen as the recipe for economic prosperity. Hence, the unquestioned adoption of this economic model in most developed economies. However, this model has been proven to be injurious to human society. The damage is especially more evident in less economically resilient countries, that have little or no capacity to tap into the global manufacturing markets. The dark side of this model is that it offers a recipe for increasing the effects of climate change that are physically majorly borne by less developed economies. The OECD (2014) explains that unfortunately, a majority of less developed economies have numerous challenges, including financial inability to shoulder the burdens of investment in critical infrastructures that could shield them from the consequences of climate change. Their rich and developed counterparts on the other hand are arguably well-placed to implement climate change mitigation projects of their choice because they have the financial means and capacity to be selective. For least developed economies, such projects can only be viable if these countries are lucky enough to secure funding from external sources, more so in terms of loans, that have been seen to be detrimental to most of these economies as they drive them to unsustainable debt cycles (UNCTAD, 2018). It becomes even worse as some of the lenders demand that the loans be guaranteed by strategic public assets, like power plants (as demonstrated in the case of ZESCO in Zambia [Laterza & Mususa, 2018]), ports and any other property assets that have a high return capacity, and where any sign of forfeiture can threaten take-overs of the asset by the loaner.

Though such loans, as noted by Arimah (2017), have earned some countries significant reliance and adaptability capacity, most of these countries have become entangled in debt cycles resulting in their fragile inability to bolster their economic footing, with the cities experiencing or being the concentrated venues for much of those physical and human economic hardships. In these instances, such cities have become less capable to provide an environment to promote activities such as Foreign Direct Investments (FDIs), tourism and major business investments that would ultimately foster the liveability status of that country or its city. The reality is that most urban dwellers in such cities have to put up with numerous challenges and are at greater risk in suffering the brunt of climate change incidences whenever they occur (Katherine et al., 2019; Kulkarni & Shafei, 2018; Ray et al., 2019; Wright, Reeves, & Huq, 2016). Worse still, the dearth of infrastructural developments in these economies is also at a greater risk of being destroyed by such climate change incidences for events including extreme weather conditions, heat waves, extreme cold events, flooding, hurricanes and strong winds amongst others.

Lately, the impacts of climate change have also recorded as being more prevalent in developed economies and through incidences that are exerting unimaginable pressures upon their urban environments with increased flooding, extreme temperatures and strong winds. Due to this, it has become evident that there needs to be a global effort to bring to an end to or at least mitigate some of the obvious climate change impacts. This is well captured in the *Paris Agreement* that encourages member parties to reduce their emissions such that by 2030 global temperatures rise would be kept at least 2 °C below pre-industrial levels (United Nations, 2017). Additionally, this *Agreement* proposes the development of a framework that should promote transparency in issues like the reporting of emissions whereby those countries with highest emitting capacities should bear much bigger responsibilities than their less polluting counterparts. The *Agreement* also encourages financial flows, especially from developed countries to developing and least developed economies, such that they can also manage to invest in infrastructures that guarantee some reliance and adaptability. If the *Agreement* is something to go by, the developing and least developed countries will also benefit from climate-safe technologies that those with the capacity to develop are encouraged to knowledge transfer these technologies to developing and least developed economies to allow them build their own capacity for resilience and adaptability (Ari & Sari, 2017).

In addition to internationally instigated action plans and agreements, developed economies, especially in Europe, have also been engaging in practices like Emission Trading Schemes (ETS) (Jaraitė-Kažukauskė & Kažukauskas, 2014). These practices have allowed fewer polluting economies to benefit from the sale of their allocated emission trading permits to higher polluters, and the proceeds from such sales being redirected towards building their own climate reliance capacities. This scheme is seen to be of particular importance as it facilitates contributions to the financing toolbox that is geared towards supporting least developed economies and less polluting economies, as well as discouraging some practices that promote emissions.

In essence, the position in this chapter is that all the practices and efforts that would help reduce emissions, as well as to help climate change vulnerable economies build own capacities, are to be encouraged and should be pursued. Therefore, the succeeding sections will build on this narrative by discussing pertinent issues.

Climate Change and Urban Liveability

From the regular erratic weather incidents globally, profiled by mass media often with apocalyptic intonations, it is undoubtedly clear that most urban areas have had their liveability status negatively impacted by these increased incidences and the flow-on effects of climate change. Chapman, Azevedo, and Prieto-Lopez (2013) report that it is not surprising that cities in this modern era are under immense pressure from extreme weather conditions including extreme temperatures, with heat temperatures rising even above 50 °C in some cities, and the cold going even below −20 °C in some cities. Similarly, increased and shifting precipitation patterns are on the rise with events like extremely heavy rainfalls and unseasonal major flood events becoming a common theme in some cities and regions. The latter are even more problematic in semi-arid and arid zones where food harbouring surface soils are rapidly flooded, eroded and transported and flash flood and ponding events become more prevalent. Prolonged droughts have increasingly been experienced in the recent past, especially following extreme temperature periods and erratic flooding incidents that negatively impact upon productive agricultural fields and farms and potable water sources. Droughts have had negative impacts upon food supply chains, with urban dwellers experiencing scarcities rendering high prices for food products and basic commodities like potable water and natural supplements. Ogden (2018) confirms that there are also incidences of new traits of vector-borne diseases and illnesses that have direct correlations with climate change events that are affecting people across the development divide. Ogden (2018) explains that these patterns are partly influenced by temperature changes, as well as parasites, bacteria and viruses that are, in one way or another, also influenced by changing climatic patterns.

The above examples are just few amongst the many when research correlates data between the impacts of climate change and urban liveability. Indeed, there have been numerous recent studies (Alfredsson et al., 2018; Sofeska, 2017) that have focused upon these two pertinent aspects related to urban areas, and it is evident that climate change events have disrupted the liveability status and well-being of many cities. Besides increasing the aforementioned challenges, it is also understood that these patterns have had significant impacts upon air quality, economic development of urban areas, infrastructural investments, sometimes regional and international policy disruptions, natural and exotic terrestrial and aquatic animal communities, and most certainly upon the very backbone of environmental sustainability.

In regard to the economies of urban areas, there is a common understanding that cities are the backbone of most economies as drivers of GDP growth. UNCTAD (2018) explains that although cities occupy only less than 3% of habitable environments, they are responsible for over 80% of the GDP in most countries regardless of their development status. Therefore, negative events triggered by climate change have substantial impacts upon economies because it is from these economies that financial support to reconstruct, treat, compensate and invest in mitigation programmes is sourced. As Forzieri et al. (2018) and Z. Allam and Jones (2019) express, an economy suffers the most when even the minutest event of climate change occurs, as most if not all sectors are highly reliant upon the prospering prevailing economic situation of the country, or region.

In regard to politics, climate change has been seen to heighten some regional and international disagreements, especially in regard to the responsibility of different countries in respect to mitigation programmes aimed at lowering emissions (Dellink, Hwang, Lanzi, & Chateau, 2017). These arguments are particularly forceful by the Pacific Island nations presently who are bearing major inundation and coastal erosions to their already low-lying islets and landscapes. Many of these nations are part of the UN-designated Small Island Developing States (SIDS). A case in point is the decision taken by the political leadership of the US to withdraw from the *Paris Agreement* citing unfair treatment. But from a political lens, all these complaints boil down to how its rivals, like China, responded when the *Paris Agreement* came calling (Dai, Zhang, & Wang, 2017).

The liveability status of cities also depends upon spatial planning and how urban managers leverage the availability of new technologies to ensure optimal consumption of resources together with minimization of emissions and other negative outputs that impact upon a city's liveability status (M. Z. Allam, 2018; Z. Allam, 2019a, 2019b, 2020a, 2020b, 2020c, 2020d, 2020e; Allam & Dhunny, 2019; Allam, Tegally, & Thondoo, 2019). On this, The World Bank (2015) concludes that some of cities' pollution-prone sectors, such as energy and transportation, also play significant roles in facilitating socio-economic progress that warrants extra attention to ensure that their positive contributions are not watered down by their roles in exacerbating climate change and the deterioration of urban air quality. It is for this reason that numerous authors (Benedek, Sebestyén, & Bartók, 2018; Bower, 2017; IRENA, 2017; Kelsey & Meckling, 2018; Punda, Capuder, Pandžić, & Delimarb, 2017; REN21, 2018) have encouraged the need to adopt a paradigm shift in the production of energy from traditional fossil fuel power plants to those that accommodate alternative and renewable energies. By doing this paradigm shift, it is believed that a considerable amount of emissions would be reduced, while at the same time the city would better manage to be able to increase their production capacities to meet increasing demands. In so doing, such efforts could stimulate human liveability health by first lowering the cost of energy, and secondly, reducing the impacts caused by emissions.

According to Alam, St-Hilaire, and Kunz (2017), encouraging the adoption of renewable sources, like solar, wind, tidal, hydro, etc., would allow even small-scale production, especially on rooftops and facades. These authors argue that this shift would permit the sharing and or selling of surplus energy to their neighbours in peer-to-peer (P2P) arrangements. Thus, these arrangements could promote alternative avenues of income generation. With these forms of energy production, local governments can also relieve some pressure of having to ensure sufficient energy supply for everyone, and thus, is able to reduce some of its overheads especially linked to distribution and maintenance in areas where locals are already producing their own energy. In return, local governments can use the savings availed by renewable energies to better improve other sectors like transportation, housing and waste management, that could ultimately enliven liveability status to greater heights. Adoption of energy-efficient means of transportation, like facilitating biking, adoption of autonomous vehicles, creating infrastructures like railways that promote the use trains, offer alternate and novel measures that can allow local governments to reduce the emissions. Carter et al. (2015) also advise that spatial planning, where compact building and mixed-use houses are encouraged, offers another way of ensuring that some activities that promote pollution and compromise the living status of the urban locals, are reduced. Carter et al. (2015) further explain that such planning allows for efficiencies in energy consumption, waste collection, reduce traffic times, encourages creation of green spaces that are all variables that promote liveability and contribute to combating climate change. In addition, with these arrangements, it is possible to integrate smart urban concepts that heavily rely upon advanced technologies, such as IoT, AI and big data, amongst others, to inform decision making on various urban issues courtesy of massive data that these technologies can collect, process, store and analyse.

There is a growing body of literature on the role of cities in fostering healthy environments and healthy people. Hence, these trends point to patterns and statistics that demonstrate higher levels of urban liveability. As briefly mentioned above, the health burdens induced by climate change and the environmental risks they incur can no longer be dismissed. The latest *Global Environment Outlook* (GEO-6) reports that 25% of all deaths in the world are caused by environmental issues that could be avoided (Gupta, 2019). Healthy environments are crucial for socio-economic viability and social equity. Human health is threatened by high levels of air pollution, one of the leading causes of death and illness, claiming seven million deaths per year. Most people are dying by breathing highly toxic outdoor air, while the remaining 10% dependent upon indoor solid fuels or kerosene for cooking, heating and lighting (Landrigan et al., 2018). Still, more than half of the world's population are exposed to bad quality water and inadequate sanitation, while badly managed environments such as weak drainage, irrigation and dam programmes continue to increase their vulnerability and exposure to infectious diseases.

The stark increase in temperatures, the advent of climate disasters, such as floods and cyclones, as well as the loss in biodiversity de-exposure to natural spaces and landscapes, all negatively affect human and environmental health. This is apparent in looking at how cities have expanded rapidly, consuming areas of rich biodiversity life in its fringes, as showcased in Figs. 3.1, 3.2, and 3.3.

Human populations are also increasingly water-stressed and are being forced to migrate towards areas that are less prone to dramatic weather changes and possessing unstable ecosystems. It is important to note that with a large focus on health, and considering the consequences of global environmental change, countries have ratified international frameworks for action after making careful appraisals of their health risks and their strategies for the distribution of health gains across different sectors and policies. Important examples include the Health in All Policies (HiAP) framework (2013), the 2030 *Sustainable Development Agenda* (2015)

Fig. 3.1 City of Sao Paulo in Brazil (Image by Jaime Spaniol)

and the *New Urban Agenda* of UN-Habitat III (2016). In particular, the operational framework of Health in All Policies (HiAP), forged during the Finnish Presidency of the European Union in 2006, the 2011 *Rio Political Declaration on Social Determinants of Health*, and the 2013 *Helsinki Statement on Health in All Policies* (Rudolph, Caplan, Ben-Moshe, & Dillon, 2013), all offer and provide pragmatic tools and pathways to use and promote different forms of impact assessments that can be applied to climate change programmes.

Health impact assessments, for instance, have gained momentum over the last decades and are currently being promoted to assess the benefits and risks of climate change policies (Dannenberg, Rogerson, & Rudolph, 2019). The benefits of assessing such policies lie in transparent and accountable decision-making processes and in facilitating synergies across non-health sectors, manufacturing industries and large development projects, in order to improve health and equity. Impact assessments provide estimates on economic savings and the long-term gains of programmes, policies and projects directly or indirectly aiming for climate change mitigation. There are a number of exposure pathways that are relevant to climate change discussions. These pathways have been extensively reviewed in literature namely: air quality, asbestos exposure, toxins, nutrition, public and green space, urban transport planning, community and

Fig. 3.2 City of Mumbai in India (Image by Abhay Singh)

social services, economic development, business food and nutrition, housing, human injury and security, car parking, physical activities and universal design, infectious diseases, social cohesion, urban design, transport, clinical and electronic waste management, water availability and quality, clinical waste, housing, salt consumption, cigarette smoking, and economic investment programmes (Thondoo, Rojas-Rueda, Gupta, de Vries, & Nieuwenhuijsen, 2019).

The potential for health impact assessments to undertake cost analyses and to provide financial estimates is valued and applied significantly to

Fig. 3.3 City of Tokyo in Japan (Image by Terence Starkey)

the exposure pathways between environmental change and health. This is of particular relevance in middle-income developed countries, where for instance, pollution-related diseases drain nearly 7% of total GDP health care costs compared to only 1.7% in high-income countries. A Health Impact Assessment (HIA) in São Paulo in Brazil demonstrated that if the city could diminish air pollution due to particulate matter (PM 2.5) by 5 g/m^3, then a total amount of $USD 4.96 billion would be saved annually in terms of São Paulo resident health costs (Abe & Miraglia, 2016). Sun, Fang, and Sun (2018) provide another relevant example in their quantitative estimations that air pollution abatement to World Health Organization standards in 13 cities would save up to 810.48 billion ¥ in 13 Chinese cities. Still in China, it is believed that by 2020 the economic development of transport in cities will cause an additional 51,000 extra hospital admissions and more than 850,000 restricted activity days (Vu, Le, Pham, & Hens, 2013).

The economic value of health and the loss of investment induced by potential environmental degradations affecting urban liveability are of increasing concern to governments and private investors. Therefore, leveraging on different health and liveability-based technologies, and impact assessments, allows for a robust link between urban liveability and strong economic outputs that are prerequisites in the fight against climate change, especially for those regions in the Global South that are more vulnerable. Technological solutions have also been seen to gain momentum in the Global North, especially as call for reductions in emissions intensifies, as these are main culprits in terms of substantial emissions production. Though most cities in wealthy economies have attained respectable levels of urban liveability, the employment of advanced smart technologies would continue to make accelerate this achievement, wherein the same has the potential to ignite tangible efforts in the Global South where urban liveability is more crucial. Thus, strategies need to justify an increased commitment in doing everything possible to attain it. Nevertheless, as will be explored in the next section, scarcity of finances to finance projects that would make them improve on liveability remains a challenge.

Leveraging on these advanced technologies allows for a health link between urban liveability and strong economic output. This link is a prerequisite in fighting climate change especially for those regions in the Global South that are always vulnerable. Technological solutions have been seen to have gained momentum in the Global North, especially as the call for reductions in emissions intensifies, because these are main culprit countries in terms of their substantial emissions production. Though most cities in wealthy economies have attained respectable levels of urban liveability, the employment of advanced smart technologies will continue to accelerate these achievements offering the potential to ignite tangible efforts in the Global South where urban liveability is crucial. Thus, there is a need to justify increased climate change commitments in doing everything possible to attain emissions reductions and climate change mediation outcomes. Nevertheless, as will be explored in the next section, scarcity of finances to finance projects that would make them improve on liveability remains a challenge.

Climate Change Funding Avenues

From a wide analysis of literature, the impacts of climate change have had devastating impacts upon every aspect of the earth's fauna and flora, its environment and habited and inhabited environments, and human-constructed infrastructures. But these impacts are perceived to be far much worse impacting upon humans who are left to bear the brunt of having to restore most of these destructions and side effects, in addition to healing the 'wounds' that such impacts inflict on them socially, economically, psychologically and even politically.

Therefore, it has become paramount to use all means necessary to address this menace, especially towards ensuring that community's livelihoods, lives, properties and living environments are safeguarded. This aim is in part achieved by building capacities for people and areas as they live in and seek to adapt and develop resilience. This, as argued by Were (2018), is achieved by engaging in infrastructural development, especially focusing upon critical infrastructures that can help mitigate climate change. It also entails encouraging those countries and economies, with such infrastructures already in place, to adopt economic models that at a minimum, promote a reduction of emissions that are in the forefront in propagating climate change. While either of these strategies has the potential to part-address climate change, they are not easily achieved. This is because both strategies require massive resources and financial capacities. Hence, the core reason why most of the greatly climate change affected regions are those comprising least developed economies is because they are overwhelmed by the expensive nature of constructing the aforementioned prerequisite infrastructures (Pan, Elzen, Höhne, Teng, & Wang, 2017).

With that realization, there are now increased calls for financial assistance in the form of financial appropriations, funds and grants that are geared towards financing projects aimed at mitigation, adaptation and resiliency (Carter et al., 2015). Most of these financing options are reserved to assist least developed economies and Small Island Developing States (SIDS) that are often at risk from the consequences of climatic change. These funds are being received from different sources, as shown in Table 3.1, including the international community via institutions like the World Bank and the UNFCCC that has already proposed a fund following the *Paris Agreement* (Pan et al., 2017). Other institutions include financial banks, like the African Development Bank (AfDB) and insurance companies that are committed towards environmental sustainability and the reduction of climate change effects. Other sources, like those mentioned above, include the sale of emission tradable permits to those countries with low-emission capacities are able to trade with high-polluter countries (Shahab, Clinch, & O'Neill, 2017, 2018). Similar to the diversity in financial sources, these funds are also diverse as showcased in Table 3.1 with the common denominator in all being that they all

Table 3.1 Notable Funds available for climate change adaptation, residence and mitigation

Type/Fund name	*Cities/Countries*	*Development status*	*Repayment*	*Noted issues*	*Key references*
Green Funds		Developed/Low income/SIDS	Partly repayable		Green Climate Fund (2015)
Special Climate Change Fund	79 countries as of 2017	Low income/SIDS	No		UNFCCC (2019b)
Adaptation Fund		Low income and SIDS	No	Technical issues, politics and limited funding	UNFCCC (2019a) and Igoe (2016)
Least Developed Countries Fund	51 countries by 2017	Low income and SIDS	No	Lack of funds	Global Environmental Facility (2019) and United Nations Committees for Development Policy
Africa Climate Change Fund	African countries	Low income	No	Limited resources Slow distribution rate	African Development Bank Group (2017)
Global Energy Efficiency and Renewable Fund (GEEREF)	All developing countries	All developing countries	No	Funds only available in the form of equity	GEEREF (2019)
Congo Basin Forest Fund	Central Africa Region	Low income	No	Limited time span	ODI (2019)

Type/Fund name	*Cities/Countries*	*Development status*	*Repayment*	*Noted issues*	*Key references*
Forest Carbon Partnership Facility	47 developing countries across Africa, Asia and Latin America	Low income/developing	No	Limited to forest activities	Forest Carbon Partnership Facility (2019)
Pilot Program for Climate Resilience (PPCR)	Global: Piloted in two regions	Low income/SIDS	No	Only 28 countries covered so far	African Development Bank (2019)

target capacity building, the creation of resilience and the promotion of strategies that could lead to reductions in emissions.

All these funds have the potential to spur economies left behind, in terms of infrastructural developments, to start their journeys towards adaptation and resilience. However, it needs to be recognized that several funds are being used as tools for geopolitical influence to support capitalistic motives of different developed economies. Linnér and Klein (2017) argue that this explains the reason why the composition of board members for some funds is skewed in favour of developed economies. Those few board members representative from 'needy' economies have little or no influence in regard to deciding how and where funds should be appropriated and or distributed. This geopolitical influence is also evident in the number of stringent requirements and demands that are tied to some of these funds and their conditional need to be met or fulfilled for a country to qualify for access to financial support.

This is not the case when it comes to the plight of developed economies that are at liberty to access financial tools without necessarily having to meet all the conditional requirements. On this, Heathwood (2017) highlights that some of these funds were recently seen to influence market trading activities to favourable support local and regional markets. That is, some of the financiers have been using their financial capabilities to win trade or investment opportunities in exchange for their support of availing monies to vulnerable economies to achieve their infrastructural investment goals. A practical scenario is the case of EXIM Bank of China that is accused of influencing trade in favour of its owner country where those seeking its financing are required to allow the construction of different infrastructure by Chinese contractors (Abegunrin & Manyeruke, 2019). On the same thread, the EXIM Bank of China requires countries in need of monies to attach one or more of its critical public assets as collateral for loans. In this case, a number of economies have risked surrendering control of their assets for a good number of years to this bank for failure to honour their loan repayment agreement. This example is recently demonstrated aptly in Zambia (Laterza & Mususa, 2018), in Djibouti (Senators, 2018), in Sri Lanka (Abi-Habib, 2018) and in Pakistan (Aamir, 2017). Such financing trends are sometimes viewed as setbacks to the main agenda of establishing financing mechanism because they increase the tension and instability between developed economies, especially those with trade interests in the less developed economies. Again, this trend is currently evident in African, South-East Asian and Pacific

Island countries. This means that when one of the competing economies manages to be the first to finance an infrastructural project in a given country, its competitors are thereupon highly likely to reject any financial support requested by this very country (Zhang, Chao, Zheng, & Huang, 2017).

These underhanded tactics, tied to some of these funds and monies, especially those advanced in the form of loans, also exacerbate the risks of economic growth slowdown or stagnation. This is because most of these funds, that would be used to support growth of a sector, are direct towards repayment of loans that are used to finance only one or a few sectors. As Gurara et al. (2017) explain, this is painful to least in developing economies that require development and growth in all of their sectors if they are to address climate change risk and impacts. The formulation of such financing options, as noted by PWC (2016), sometimes promotes cases of financial impropriety including corruption, kickbacks and misappropriation. These latter issues all aid in worsening a country's debt situation, and in extension, the vulnerability of some of these economies that in reality need urgent assistance if their communities, properties and their environments are to be quality safeguarded from the wrath of climate change (Flynn, Rao, Horner, & Gashi, 2018).

The Unsustainability of Debt Financing and Urban Liveability

From the above literature, it has been established that only a few countries have the financial capacity to comfortably finance climate change mitigation programmes from their government earnings. Most countries, both in developed, developing and least developing countries, therefore have to rely upon other sources like grants, private sponsorships or through debt financing. In this case, debt financing has been seen as the most preferred and successful source of finance in many countries. Nevertheless, while its success rate has been lauded, it has also received equal measures of criticism for its potential to plunge economies into stifling long-term economic obligations, especially in respect to repayment commitments (UNCTAD, 2017). This scenario has been very real in most developing and least developing economies, where some, as noted in a report by the US (Senators, 2018), are already pressured for repayments by external debtors who eye the potentiality of obtaining unfettered access to their vital public assets that they have tabled as collateral. Such risks have been witnessed in prompting the adoption of stringent measures, both by lenders and loanees in ensuring that repayments materialize. PWC (2016) argues that while taking such precautionary measures is novel, it opens the loanees' economies to serious repercussions that compromise their urban liveability capacities, especially when these obligations are viewed from a socio-economic perspective.

With that background, it is safe to argue that issues of urban liveability, in the wake of climate change, arouse an interesting dichotomy. That is, the liveability aspect is seen to be affected by both the effects of climate change and debt financing models. Contextually, most developing economies are increasingly seen as being accustomed in their debt financing bidding to support climate change mitigation programmes and projects (Allam & Jones, 2019; Kling, Lo, Murinde, & Volz, 2018). This is worrying trend in loan financing. Such scenarios then lead to questions of whether urban liveability status is ultimately doomed, and whether traditional economic policies lead to urban problems are carved testaments in stone. On this, the position in this book is that there are other economic options that can be explored by cities and economies to mitigate the impacts of climate change without adding to their extreme debt burdens. Here, it is believed that there need to be some soul searching by those at the helm of urban leadership in regard to how they would like to finance climate change mitigation programmes, such that, as they do this, they do not jeopardize their economies. Therefore, the question of how much debt such economies can shoulder without crossing the bad debt line becomes paramount as most of these countries are already at risk from debt burdens and are being seen to acquire more debts than what their internal economies can sustain (UNCTAD, 2017). The formulation of loan instruments by some of lenders are thus seen as being very attractive on the face value for most economies to resist, but, scratching the surface of those incentives reveals a plethora of challenges ranging from collaterals, high interest rates, and unfair balances of trade to name a few (Deloitte, 2013). Such loans are seen to favour the lenders, while the loanees are entrapped in debt cycles, that not only compromise their urban liveability status, but also the vital infrastructures that they table as securities for these loans (Were, 2018). Such loans are more predominantly synonymous with China through its *Belt and Road Initiative*, where least developed economies are concerned (as discussed in different sections of this book), where they are on the verge of losing their critical infrastructures due to their failure to honour loan repayment agreements.

Another issue, that raises scepticism in debt financing, is how monies received are utilized at the project implementation phase. In developed economies, it can be seen that investments yield results, but in a majority of least developed economies, a sizeable percentage of such loans end up in the coffers of politicians, urban managers and other interested groups. Lemma Tesfaye (2015) explains that cultural vices, including corruption, misappropriation, kickbacks and diversion of monies, are rampant practices in most needy economies. Unfortunately, when such negatives abound, it is believed that even the quality of work undertaken on projects is questionable as most of those entrusted with the responsibility of quality assurance have a price tag and are willing to compromise as long as they are able to secure kickbacks from contractors and stakeholders involved in the projects. When such issues are analysed, they then raise the concern of whether it is ethical to opt for debt financing for climate change in such economies when it is true that not every penny will be committed to the target project.

While debt financing for climate change is important for least developed economies, it should not be at the risk of succumbing to internal country corruption practices and a failure to focus upon the bigger picture that seeks to support vulnerable economies from establishing critical infrastructures vital to their survival. Therefore, as is proposed by the International Monetary Fund (IMF), it is paramount that debt financing be intertwined with transparency and anti-corruption measures that include monitoring mechanisms and accountability (Wei, 2001). This, however, does not imply that corruption will be absent from the equation, but this will aid in its dissipation. Such measures could also involve support from international bodies and institutions for devising more locally calibrated financial processes to ensure that resources and monies suspected to be obtained from scrupulous ways are not staked in foreign accounts.

In addition to those internal ethical and moral concerns, externally, it is argued that developed economies owe the poor countries ‘climate debt’ because they are highly responsible for most of the climate change events due to their economic growth models that perpetuate unsustainable levels of emissions. Pickering and Barry (2012) posit that wealthy countries are answerable to issues of climate change because most of their wealth is directly or indirectly tied to activities that have for many years contributed to human-induced climate change effects. Additionally, their inactions to combat or slow down the emissions have driven a majority of the less developed economies into further poverty perpetuated by climate change events.

In an ideal situation devoid of political interference and dishonesty, wealthy economies should compensate less developed ones, but as it stands now, wealthy economies are competing for the little available funds for climate change mitigation (Yeo, 2019). Further, as noted in the previous section, wealthy economies have an upper hand since the composition of such fund boards is skewed in their favour. In support of this assertion, Asayama and Hulme (2019) highlight that debt financing of climate change mitigation programmes, has the potential to give high polluters the leeway to continue with their actions because they argue that they are contributing towards these climate funds. Consequently, with each economy experiencing some levels of mitigation investment, high polluters may also relax in their abatement, and thereupon reduce the rate at which emissions reductions is envisioned in the *Paris Agreement*. This reduction translates to an infinite time period by which they argue that the world is now safe from climate change, at the expense of less developed economies and SIDS economies, that need to continue absorbing more debt to strengthen their adaptability.

On a different but related issue, Dahlmann, Branicki, and Brammer (2019) express another ethical concern that crowds debt financing. According to them, due to pressure from the international community demanding a reduction in emissions and subsequent reporting upon emissions target after a given period, most companies and corporations, within some jurisdictions have been observed as engaging in some dishonest practices, including 'greenwashing', while in reality their contribution to emissions is unabated. In such circumstances, vulnerable economies and their populations are the one on the receiving end, as they have to bear the costs of climate change from such unethical practices. At the same time, these vulnerable economies and their populations are made to purchase products from companies and corporations engaging in such practices at relatively higher prices (Gatti, Seele, & Rademacher, 2019). On this, Chen, Huang, Wang, and Chen (2018) argue that climate change has given some companies and corporations the impetus to engage in Public Relations (PR) and branding strategies, thus exacerbating and disguising the climate debt financing crisis.

Following these concerns on debt financing, it is the position of this book that other potential financing avenues should be explored. As will be expounded further in Chapter 4, such strategies, like economic incentivization of local climate change mitigation programmes and local capacity building, are just a few of them. An example of such a strategy is the UNDP's *Community Water Initiative* that has been operating in different parts of the globe by providing water projects to the most needy in least developed economies (UNDP, 2010). When pursued, these strategies have the potential to stimulate the development of more sustainable and liveable urban fabrics and should be encouraged as they can offset some of the challenges of relying upon debt financing.

References

Aamir, A. (2017, December 27). *Will Gwadar go the way of Hambantota? Why Chinese loans to Pakistan are sparking takeover fears along the economic corridor*. Retrieved from https://www.scmp.com/comment/insight-opinion/article/2125783/will-gwadar-go-way-hambantota-why-chinese-loans-pakistan-are.

Abe, K., & Miraglia, S. (2016). Health impact assessment of air pollution in São Paulo, Brazil. *International Journal of Environmental Research and Public Health, 13*(7), 694.

Abegunrin, O., & Manyeruke, C. (2019). *China's power in Africa: A new global order*. Cham: Springer.

Abi-Habib, M. (2018, June 25). How China got Sri Lanka to cough up a port. *New York Times*.

African Development Bank. (2019). *Pilot Program for Climate Resilience (PPCR)*. Retrieved from https://www.afdb.org/en/topics-and-sectors/initiatives-partnerships/climate-investment-funds-cif/strategic-climate-fund/pilot-program-for-climate-resilience-ppcr/.

African Development Bank Group. (2017, July 10). *Africa Climate Change Fund launces second call for proposals*. Retrieved from https://www.afdb.org/en/news-and-events/africa-climate-change-fund-launches-second-call-for-proposals-17169/.

Alam, M., St-Hilaire, M., & Kunz, T. (2017). An optimal P2P energy trading model for smart homes in the smart grid. *Energy Efficiency, 10*(6), 1475–1493.

Alfredsson, E., Bengtsson, M., Brwon, H. S., Isenhour, C., Lorek, S., & Stevis, D. (2018). Why achieving the Paris Agreement requires reduced overall consumption and production. *Sustainability: Science, Practice and Policy, 14*(1), 1–5.

Allam, M. Z. (2018). *Redefining the smart city: Culture, metabolism and governance*. Case Study of Port Louis, Mauritius (PhD). Curtin University, Perth, Australia. Retrieved from https://espace.curtin.edu.au/handle/20.500.11937/70707.

Allam, Z. (2019a). Achieving neuroplasticity in artificial neural networks through smart cities. *Smart Cities, 2*(2), 118–134.

Allam, Z. (2019b). The emergence of anti-privacy and control at the Nexus between the concepts of safe city and smart city. *Smart Cities, 2*(1), 96–105.

Allam, Z. (2020a). Data as the new driving gears of urbanization. In Z. Allam (Ed.), *Cities and the digital revolution: Aligning technology and humanity* (pp. 1–29). Cham: Springer.

Allam, Z. (2020b). Digital urban networks and social media. In Z. Allam (Ed.), *Cities and the digital revolution: Aligning technology and humanity* (pp. 61–83). Cham: Springer.

Allam, Z. (2020c). On culture, technology and global cities. In Z. Allam (Ed.), *Cities and the digital revolution: Aligning technology and humanity* (pp. 107–124). Cham: Springer.

Allam, Z. (2020d). Privatization and privacy in the digital city. In Z. Allam (Ed.), *Cities and the digital revolution: Aligning technology and humanity* (pp. 85–106). Cham: Springer.

Allam, Z. (2020e). Urban Chaos and the AI Messiah. In Z. Allam (Ed.), *Cities and the Digital Revolution: Aligning technology and humanity* (pp. 31–60). Cham: Springer.

Allam, Z., & Dhunny, Z. A. (2019). On big data, artificial intelligence and smart cities. *Cities, 89*, 80–91.

Allam, Z., & Jones, D. (2019). Climate change and economic resilience through urban and cultural heritage: The case of emerging small island developing states economies. *Economies, 7*(2), 62.

Allam, Z., Tegally, H., & Thondoo, M. (2019). Redefining the use of big data in urban health for increased liveability in smart cities. *Smart Cities, 2*(2), 266–278.

Ari, I., & Sari, R. (2017). Differentiation of developed and developing countries for the Paris Agreement. *Energy Strategy Reviews, 18*, 175–182.

Arimah, B. (2017). Infrastructure as a catalyst for the prosperity of African Cities. *Procedia Engineering, 198*, 245–266.

Asayama, S., & Hulme, M. (2019). Engineering climate debt: Temperature overshoot and peak-shaving as risky subprime mortgage lending. *Climate Policy, 19*(8), 937–946.

Benedek, J., Sebestyén, T.-T., & Bartók, B. (2018). Evaluation of renewable energy sources in peripheral areas and renewable energy-based rural development. *Renewable and Sustainable Energy Reviews, 90*, 516–535.

Bower, F. (2017). The long-term effect of state renewable energy incentive programs. *Journal of Environmental and Resource Economics at Colby, 4*(1), 1–13.
Carter, J. G., Cavan, G., Connelly, A., Guy, S., Handley, J., & Kazmierczak, A. (2015). Climate change and the city: Building capacity for urban adaptation. *Progress in Planning, 95*, 1–66.
Chapman, L., Azevedo, J. A., & Prieto-Lopez, T. (2013). Urban heat & critical infrastructure networks: A viewpoint. *Urban Climate, 3*, 7–12.
Chen, Y.-S., Huang, A. F., Wang, T.-Y., & Chen, Y.-R. (2018). Greenwash and green purchase behaviour: The mediation of green brand image and green brand loyalty. *Total Quality Management & Business Excellence, 29*(1), 1–16.
Covert, T., Greenstone, M., & Knittel, C. R. (2016). Will we ever stop using fossil fuels? *The Journal of Economic Perspectives, 30*(1), 117–137.
Dahlmann, F., Branicki, L., & Brammer, S. J. J. O. B. E. (2019). Managing carbon aspirations: The influence of corporate climate change targets on environmental performance. *Journal of Business Ethics, 158*(1), 1–24.
Dai, H.-C., Zhang, H.-B., & Wang, W.-T. (2017). The impacts of U.S. withdrawal from the Paris Agreement on the carbon emission space and mitigation cost of China, EU, and Japan under the constraints of the global carbon emission space. *Advances in Climate Change Research, 8*(4), 226–234.
Dannenberg, A. L., Rogerson, B., & Rudolph, L. (2019). Optimizing the health benefits of climate change policies using health impact assessment. *Journal of Public Health Policy*, 1–16. https://doi.org/10.1057/s41271-019-00189-y.
Dellink, R., Hwang, H., Lanzi, E., & Chateau, J. (2017). *International trade consequences of climate change*. Paris. Retrieved from https://www.oecd-ilibrary.org/content/paper/9f446180-en.
Deloitte. (2013). *Funding options: Alternative financing for infrastructure development*. Sydney, Australia. Retrieved from https://www2.deloitte.com/content/dam/Deloitte/au/Documents/public-sector/deloitte-au-ps-funding-options-alternative-financing-infrastructure-development-170914.pdf.
Flynn, M., Rao, A. K., Horner, J., & Gashi, D. S. (2018). *Smart cities funding and financing in developing economies: Assisting development cities to finance their infrastructure gap through private sector participation apporaches*.
Forest Carbon Partnership Facility. (2019). *The Forest Carbon Partnership Facility (FCPF)*. Retrieved from https://www.forestcarbonpartnership.org/.
Forzieri, G., Bianchi, A., Silva, F. B. E., Marin Herrera, M. A., Leblois, A., Lavalle, C., … Feyen, L. (2018). Escalating impacts of climate extremes on critical infrastructures in Europe. *Global Environmental Change, 48*, 97–107.
Gatti, L., Seele, P., & Rademacher, L. (2019). Grey zone in—Greenwash out: A review of greenwashing research and implications for the voluntary-mandatory transition of CSR. *International Journal of Corporate Social Responsibility, 4*(1), 6.

GEEREF. (2019). Bringing clean power to developing countries: Fighting climate change with equity investments. Retrieved from https://geeref.com/.

Global Environmental Facility. (2019). *Least Developed Countries Fund—LDCF.* Retrieved from https://www.thegef.org/topics/least-developed-countries-fund-ldcf.

Green Climate Fund. n.d. *The Climate Challenge.* Retrieved from https://www.greenclimate.fund/home.

Green Climate Fund. (2015). *Financial terms and conditions of the fund's instruments.* Songdo: Republic of Korea.

Gupta, J. (2019). *What does UN environment's GEO-6 mean for INEA?* Cham: Springer.

Gurara, D., Klyuev, V., Mwase, N., Presbitero, A., Xu, X. C., & Bannister, G. (2017). *Trends and challenges in infrastructure investment in low-income developing countries.* Washington, DC: International Monetary Fund.

Heathwood, A. (2017, January 11). *The unintended consequences of UNESCO world heritage listing.* Retrieved from https://theconversation.com/the-unintended-consequences-of-unesco-world-heritage-listing-71047.

Henriques, I., & Sadorsky, P. (2018). Investor implications of divesting from fossil fuels. *Global Finance Journal, 38,* 30–44.

Igoe, M. (2016, November 18). *Who's fighting for the Adaptation Fund?* Retrieved from https://www.devex.com/news/who-s-fighting-for-the-adaptation-fund-89189.

IPCC. (1997). *The regional impacts of climate change: An assessment of vulnerability* (R. T. Watson, M. C. Zinyowera, & R. H. Moss Eds.). Cambridge, UK: Cambridge University Press.

IRENA. (2017). *Rethinking energy 2017.* Abu Dhabi: International Renewable Energy Agency.

Jaraite˙-Kazˇukauske˙, J., & Kazˇukauskas, A. (2014). Do transaction costs influence firm trading behavior in the European emission trading system? *Environmental and Resource Economics, 62*(3), 583–613.

Johnsson, F., Kjärstad, J., & Rootzén, J. (2018). The threat to climate change mitigation posed by the abundance of fossil fuels. *Climate Policy, 19*(2), 258–274.

Katherine, G., Paul, Y., Robert, W., Ebenezer, A., Mercy, A., Samuel, C., … Cuthbert, N. (2019). *Vulnerability to extreme weather events in cities: Implications for infrastructure and livelihoods.* Loughborough University, Loughborough.

Kelsey, N., & Meckling, J. (2018). Who wins in renewable energy? Evidence from Europe and the United States. *Energy Research & Social Science, 37,* 65–73.

Kling, G., Lo, Y., Murinde, V., & Volz, U. (2018). Climate vulnerability and the cost of debt. *SSRN Electronic Journal, 3198093,* 1–30.

Kulkarni, A. R., & Shafei, B. (2018). Impact of extreme events on transportation infrastructure in Iowa: A Bayesian network approach. *Transportation Research Record, 2672*(48), 45–57.

Landrigan, P. J., Fuller, R., Acosta, N. J. R., Adeyi, O., Arnold, R., Baldé, A. B., … Breysse, P. N. (2018). The Lancet Commission on pollution and health. *The Lancet, 391*(10119), 462–512.

Lanfranchi, G., Herrero, A. C., Palenzuela, S. R., Camilloni, I., & Bauer, S. (2018). *The new urban paradigm* (Economics Discussion Papers No 2018-70). Retrieved from http://www.economics-ejournal.org/economics/discussionpapers/2018-70.

Laterza, V., & Mususa, P. (2018, October 11). *Is China really to blame for Zambia's debt problems?* Retrieved from https://www.aljazeera.com/indepth/opinion/china-blame-zambia-debt-problems-181009140625090.html.

Lemma Tesfaye, T. (2015). Corruption, debt financing and corporate ownership. *Journal of Economic Studies, 42*(3), 433–461.

Linnér, B.-O., & Klein, R. J. T. (2017). Climate change adaptation and the Least Developed Countries Fund (LDCF): Qualitative insights from policy implementation in the Asia-Pacific. *Climatic Change, 140*(2), 209–226.

ODI. (2019). *Congo Basin Forest Fund.* Retrieved from https://climatefundsupdate.org/the-funds/congo-basin-forest-fund/.

OECD. (2014). *Cities and Climate Change: National governments enabling local action.* OECD Policy Perspectives.

Ogden, L. E. (2018). Climate change, pathogens, and people: The challenges of monitoring a moving target. *BioScience, 68*(10), 733–739.

Pan, X., Elzen, M. D., Höhne, N., Teng, F., & Wang, L. (2017). Exploring fair and ambitious mitigation contributions under the Paris Agreement goals. *Environmental Science & Policy, 74*, 49–56.

Pickering, J., & Barry, C. (2012). On the concept of climate debt: Its moral and political value. *Critical Review of International Social and Political Philosophy, 15*(5), 667–685.

Punda, L., Capuder, T., Pandžić, H., & Delimarb, M. (2017). Integration of renewable energy sources in southeast Europe: A review of incentive mechanisms and feasibility of investments. *Renewable and Sustainable Energy Reviews, 71*, 77–88.

PWC. (2016). *A Comprehensive analysis and review of investment, production and export incentives in Malawi.* Blantyre, Malawi: PWC.

Ray, D. K., West, P. C., Clark, M., Gerber, J. S., Prishchepov, A. V., & Chatterjee, S. (2019). Climate change has likely already affected global food production. *PLoS ONE, 14*(5), e0217148.

REN21. (2018). *Renewables 2018 Global Status Report.* Paris, France.

Rudolph, L., Caplan, J., Ben-Moshe, K., & Dillon, L. (2013). *Health in all policies: A guide for state and local governments.* Washington, DC: American Public Health Association.

Senators. (2018). *US Senate China IMF Letter.* Washington, DC.

Shahab, S., Clinch, J. P., & O'Neill, E. (2017). Impact-based planning evaluation: Advancing normative criteria for policy analysis. *Environment and Planning B: Urban Analytics and City Science, 46*(3), 534–550.

Shahab, S., Clinch, J. P., & O'Neill, E. (2018). Accounting for transaction costs in planning policy evaluation. *Land Use Policy, 70,* 263–272.

Sofeska, E. (2017). Understanding the livability in a city through smart solutions and urban planning toward developing sustainable livable future of the city of Skopje. *Procedia Environmental Sciences, 37,* 442–453.

Sun, D., Fang, J., & Sun, J. (2018). Health-related benefits of air quality improvement from coal control in China: Evidence from the Jing-Jin-Ji region. *Resources, Conservation and Recycling, 129,* 416–423.

The World Bank. (2015). *World Inclusive Cities Approach Paper.* MA: Danver.

Thondoo, M., Rojas-Rueda, D., Gupta, J., de Vries, D. H., & Nieuwenhuijsen, M. J. (2019). Systematic literature review of health impact assessments in low and middle-income countries. *International Journal of Environmental Research and Public Health, 16*(11), 2018.

UN Environment. (2019). *Cities and climate change.* Retrieved from https://www.unenvironment.org/explore-topics/resource-efficiency/what-we-do/cities/cities-and-climate-change.

UNCTAD. (2017). *Debt vulnerability in developing countries: A new debt trap?* New York and Geneva. Retrieved from https://unctad.org/en/PublicationsLibrary/gdsmdp2017d4v1_en.pdf.

UNCTAD. (2018). *Scaling up finance for the sustainable development goals: Experimenting with models of multilateral development banking.* New York and Geneva: UNCTAD.

UNDP. (2010). *UNDP community water intiative: Fostering water security and climate change adaptation and mitigation.* New York, NY: United Nations Development Programme (UNDP).

UNFCCC. (2019a). *Adaptation Fund.* Retrieved from https://unfccc.int/process/bodies/funds-and-financial-entities/adaptation-fund.

UNFCCC. (2019b). *The Special Climate Change Fund (SCCF).* Retrieved from https://unfccc.int/topics/climate-finance/resources/reports-of-the-special-climate-change-fund.

United Nations. (2017). *Urban environment related mitigation benefits and cobenefits of policies, practices and actions for enhancing mitigation ambition and options for supporting their implementation.* Retrieved from https://unfccc.int/sites/default/files/resource/docs/2017/tp/02.pdf.

United Nations Committees for Development Policy. n.d. *Least Development Countries Fund (LDCF)*. Retrieved from https://www.un.org/ldcportal/least-developed-countries-fund-ldcf/.

Vu, V.-H., Le, X.-Q., Pham, N.-H., & Hens, L. (2013). Application of GIS and modelling in health risk assessment for urban road mobility. *Environmental Science and Pollution Research, 20*(8), 5138–5149.

Wei, S.-J. (2001, April 30). *Corruption and globalization*. Retrieved from https://www.brookings.edu/research/corruption-and-globalization/.

Were, A. (2018). Debt trap? Chinese loans and Africa's development options. *South African Institute of International Affairs Policy Insights, 66*, 1–13.

Wright, H., Reeves, J., & Huq, S. (2016). *Impact of climate change on least developed countries: Are the SDGs possible?*

Yeo, S. (2019). Where climate cash is flowing and why it's not enough. *Nature, 573*(7774), 328–331.

Zhang, Y.-X., Chao, Q.-C., Zheng, Q.-H., & Huang, L. (2017). The withdrawal of the U.S. from the Paris Agreement and its impact on global climate change governance. *Advances in Climate Change Research, 8*(4), 213–219.

CHAPTER 4

Economically Incentivizing Urban Sustainability and Resilience

Abstract The impacts of climate change on cities are causing disruptions to urban life resulting in negative affecting economic outputs. Mitigation plans to protect vital urban infrastructures as well as the safeguarding the integrity, efficiency and performance of urban economies however tend to neglect countries that may need it the most, for example, Small Island Developing States (SIDS) and low-income economies. Void of financial means to invest in expensive mitigation projects that are essential to their survival, these countries are turning towards foreign financial aid and loans that impose unsustainable debt cycles upon their economies that ultimately impact upon the liveability levels of their urban fabric. This chapter presents a new model for catalysing foreign and local investment in projects essential towards the achievement of urban resilience while catering for UNESCO's *Sustainable Development Goal 11* and the *New Urban Agenda*. This chapter offers policy makers and urban economists an alternate avenue for rejuvenating urban economies in low-income and SIDS countries over short and medium terms through the effective change in local policies and legislation.

Keywords Sustainability · Resilience · Cities · Debt · Fiscal incentives · Urban economics

Z. Allam et al., *Cities and Climate Change*,
Palgrave Studies in Climate Resilient Societies,
https://doi.org/10.1007/978-3-030-40727-8_4

Introduction

The world is witnessing unparalleled urbanization rates driven by an upsurge in urban population growth that the World Bank (2010) believes will reach beyond 70% by the year 2050. With the right strategies, this urbanization growth can be harnessed to enable numerous positive benefits to cities, countries and even regions. Such benefits can include increasing economic growth, the creation of job opportunities, and, where managed properly, the improvement of liveability and societal quality levels (M. Z. Allam, 2018; Allam, 2012; Allam & Jones, 2018b; Allam & Newman, 2018a, 2018b; Siew & Allam, 2017).

This conclusion is further supported by Palanivel (2017) and the UNDP (2016) who demonstrate that most urbanized countries tend to be the richest with a higher liveability index. However, on the other side, researchers highlight that urbanization may also have far-reaching impacts potentially reversing all gains that may be associated with urbanization especially in developing, low-income economies and Small Island Developing States (SIDS). This conclusion is applicable because most cities, borne as a result of urbanization in these economies, are not well planned, most lack basic infrastructure like considered transport systems (Palei, 2015), quality sewerage and water supply systems (Miller & Hutchins, 2017), viable sustainable energy sources, credible waste management infrastructures, robust educational facilities (Srinivasu & Rao, 2013) and quality health infrastructure (Aliyu & Amadu, 2017). In these economies, it is evident that municipal councils and local governments lack the financial capacity to address these challenges because the majority of economic activities in some cities are still informal. Hence, there is no direct revenue being gathered by the public sector, and in some cases this revenue amounts to no gains (UN-Habitats, 2015a). These challenges are also compounded by governance and lay-level corruption, unsatisfactory land control systems, widespread informal settlements and the high cost of land (Shami & Majid, 2014).

It is however believed that these challenges are surmountable if there is a speedy establishment of new, and improvements to, existing infrastructures as demonstrated in recent changes to the Singaporean and Malaysian economies. Both these economies have dramatically emerged from similar challenges and triumphed (Asian Development Bank, 2014; Clements-Croome, Marson, Yang, & Airaksinen, 2017).

The UNDP report (UNDP, 2016) and Palanivel (2017) both posit that the challenges facing cities could be addressed by alleviating poverty, improving infrastructure and curbing pollution. Parker and Simpson (2018) further acknowledge that robust infrastructure within a city can foster its economic attractiveness, thus boosting investor confidence and ultimately helping spur economic growth both at macro- and micro-levels. Infrastructure investment is thus the key to economic growth of a city. Kodongo and Ojah (2016) affirm this conclusion in their analysis of economic growth in the sub-Saharan region of Africa demonstrating that it is, and has been, directly influenced by the level of infrastructure development. In this sub-Saharan context, with an improved economy, it becomes much easier to address challenges such as poverty, sustainability and pollution. From these, one can garner that the improvement of infrastructure should be emphasized, and ensuring that investments are geared towards sustainability, safety, inclusivity and resilience, as narrated in *Sustainable Development Goal 11* (United Nations, 2018).

In view of current local challenges, a majority of SIDS and low-income countries are embracing infrastructural development with the aim of improving their economic landscape. Borrowing from the extensive analysis by Kodongo and Ojah (2016), it is seen that when cleverly planned and implemented, those developments have been proven to spur economic growth. This is true when one appreciates that different infrastructure installations and strategies are integral to production, apart from supporting human resources and financial capital. This argument is also supported by Shi, Guo, and Sun (2017) who demonstrate how China's economic growth is, by a larger percentage, supported by infrastructure investment articulated in China's National Council policy agenda adopted in the 1980s. Similar results, of extensive economic growth, have been hosted by other Asian countries who have embraced this strategy of rapid infrastructure development and investment. For instance, Malaysia (Bakar & Mat, 2017), Singapore (Clements-Croome et al., 2017) and Korea (Asian Development Bank, 2014) are leading economies courtesy of their infrastructure investments and associated developments. This is an economic position that they respectively entered into some decades back. Even though the sustainability of Dubai in the United Arab Emirates (UAE) is disputed (Nadali, Thomas, & Taleb, 2015), whereby the City of Dubai is the beneficiary of robust infrastructure investment made by its government that has contributed to the City being a destination of choice for investors and tourists (Arafat, Bing, & Al-Mutawakel, 2018; Zaidan & Kovacs, 2017).

The notable commonality in these economies is their emphasis upon improving their basic infrastructures including transportation, telecommunication, energy, water and sewerage (Shi et al., 2017).

Investing in rapid infrastructure development is an expensive expedition both in terms of financial and resource inputs (Schmidt-Traub & Sachs, 2015). This hindrance is evident even though new technologies, like artificial intelligence and blockchain, can assist in providing more efficient and cheaper efforts (Allam, 2018b; Allam & Dhunny, 2019). Within this context, investments still require a concerted effort and contribution from all quarters of the host economic sectors and industries, and this is a pointer of the challenges that SIDS and low-income economies have been experiencing. From their lack of finance and expertise, existing SIDS infrastructures are being haphazardly built at the same time as inherently supporting the proliferation of informal sectors, non-renewable energy generation activities and poor waste management practices (Africa Development Bank, 2018). This trend has exposed these economies to a myriad of challenges that are being compounded by the impacts of climate change. Most of these countries, due to disruptions in their weather patterns, have experienced flash floods from excessive surface water runoffs, significant water-related erosion effects and ponding, are which are leading to a destruction of various infrastructures, properties and sometimes loss of life (Dong et al., 2018; Lafortezza, Chen, Van den Bosch, & Randrup, 2018). This is occurring without discussions about the comparable impacts that are occurring from irregular seawater and sea-level changes and fluctuations, and irregular climatic events.

The expansion of cities has also resulted in a reduction in the supply of resources such as potable water and food, and even administrative services, due to a lack of funds and organizational incapability to deal with these larger systems. In contrast, these challenges are not evident in developed countries, especially those that have embraced the concept of smart cities. In such places, developed countries have successfully incorporated optimal use of resources, allowed for and enabled environmental sustainability opportunities, and allowed for the construction of mixed-use mega-structures that accommodate many people utilizing only a reasonable space or developable footprint.

The concept of smart cities is however not necessarily the most appropriate model for SIDS and low-income economies as it necessitates investments on a large scale with the establishment of complex infrastructures while enhancing existing ones (Allam, 2018a; Allam & Newman, 2018a). In view of the need for infrastructure investment, the economic strain placed upon SIDS and low-income economies is a challenge. Solving this challenge by seeking finance from external, often foreign government sources, has driven these SIDS and low-income economies into short-, medium- and or long-term debt traps (Payer, 1975), hence, risking the loss of their country's strategic assets and a decrease in liveability levels.

This is not to say, as embodied in the main narrative of this paper that physical infrastructure is an important component of development, or that establishing a Special Economic Zone (SEZ) may well be superior to relying on internal public/private partnerships or foreign loans from abroad. Rather, it is very evident in SIDS economies that medium-large-scale infrastructure (airports, ports, tourism facilities, etc.) investments can be instrumental where physical ocean isolation is a key characteristic.

In this regard, the authors acknowledge that SEZs have a decades-long history around the world, with very mixed results, and that such results cannot be attributed alone to infrastructure project investment. In making this acknowledgement, it is however evident that while SEZs have played a role in the successes of the Mauritius and Singapore economies, several other (social, economic, resource quality, tourism aesthetic, transit interchange locational) policy factors to some extent have played critical and or complex roles in creating these 'economic miracles'.

This chapter explores this issue in depth, reflecting upon Indian Ocean SIDS economies specifically, and SIDS economies generally, in the following sections.

Background

Low-income economies and SIDS have for long been contending with numerous challenges especially due to their meagre infrastructure development and poor economic bases. These challenges have been compounded by the wide-ranging impacts of climate change upon their cities. Additionally, all of these cities are port cities meaning that they are all directly physically exposed to sea-related climate change impacts. These wide-ranging impacts of climate change are more pronounced in some cities due to no, or a lack of, mitigation programmes or strategies being in place or being implemented (OECD, 2014a). Natural disasters have heavily impacted upon these economies, just like they affect cities in developed countries; though not with the same micro-level and exponential impacts (Hughes & Sarzynski, 2015). The impacts of climate change have been diverse and have often directly impacted upon communities whom struggle economically. For example, one result of the impact of climate change is the stress being placed upon the provision of clean potable water and availability of a sufficient food supply that is often seen lacking in informal settlements and slums (Burton et al., 2013; United Nations, 2016). An increase in mainstream and seasonal tropical diseases is additionally contributing to the pressure and complexity of the health of these communities, affecting the productivity of workers and their respective economy at large. These patterns arise from increased temperatures (Emilsson & Sang, 2017), consumption of contaminated and unsafe water and poor nutrition due to low-quality food supplies (Burton et al., 2013). From an urban science viewpoint, climate change is also partly responsible for urban sprawl and the spread of informal settlements and slums in most developing countries and SIDS. These patterns have been occurring due to human displacements by flood, bushfires (Hales et al., 2007; McDonald, 2017) and the deterioration of agricultural lands (Burton et al., 2013) forcing human migrations to urban centres where basic infrastructures are reputedly well catered for.

In an OECD (2014a) report, it is demonstrated how large amount of resources, both in terms of finance, human capital and resources and assets, are utilized in addressing the impacts of climate change. With mitigation strategies in place, as is emphasized in the UN's *Sustainable Development Goals* (SDGs) (UN, 2015) and the *New Urban Agenda* (United Nations, 2016), such resources can be used towards the improvement of infrastructures, upgrading the quality of human settlements and in improving human social amenities. The *New Urban Agenda* policy articulates that cities ensure the inclusion of sustainability principles in development plans by supporting *SDG 11* that calls for 'inclusive, safe, resilient and sustainable' cities in various levels of urban planning, urban management and governance, and design activities. This is of particular interest to low-income countries and SIDS so that they can match their counterparts, in developing countries, in both infrastructure and economy (Van Noorloos & Kloosterboer, 2018).

From the literature, there are numerous indicators that SIDS and low-income economies have been striving to implement. These include measures and policies, to align to *SDGs* and the *New Urban Agenda* as advanced by *UN-Habitat* (or, the *United Nations Human Settlements Programme*), and supported by the United Nations (Bah, Faye, & Geh, 2018; Lee, 2014; Sepasgozar, Hawken, Sargolzaei, & Foroozanfa, 2018; Smit, Musango, Kovacic, & Brent, 2017). There is also evidence of the adoption of technologies, such as the smart cities concept (Capdevila & Zarlenga, 2015; Eko Atlantic, n.d.; Steadman, 2013), the use of big data, artificial intelligence (AI), The Internet of Things (IoT) and blockchain technologies to solve persistent problems facing cities (M. Z. Allam, 2018; Allam, 2014, 2018a, 2018c; Allam & Jones, 2018a, 2018b; Allam & Newman, 2018b; Lim, Kim, & Maglio, 2018; Pioletti, 2016).

Nevertheless, it is evident that the main shortcomings that confront the majority of these cities, especially in low-income economies, are financial constraints and a dearth of technical expertise (OECD, 2014a). These reasons have driven respective country policy pursuit for financial support both internally and externally (Attapattu & Padmasiri, 2018; Farquharson, Mästle, Yescombe, & Encinas, 2011). In particular, due to their small economies, some countries do not have the capacity to borrow through local banks and institutions and have therefore been turning towards external foreign borrowing. Luckily, there are numerous financial institutions and economies that offer such support at what are believed to being offered at reasonable terms, as noted by Ehlers (2014) and the World Bank Group (2014). The ability to borrow and the willingness of lenders to 'extend a helping hand' have however not been beneficial to all countries. This is because some countries face complex terrains in enabling debt repayments, and such repayments can be internally marred by corruption, instable political governance, religious instability, poor debt management policies and weak governance models (UNCTAD, 2017, 2018). A case in point is Sri Lanka, as documented by Weerakoon (2017), that was forced to forfeit its port for a record 99 years to China to repay an infrastructure development loan. A similar tragedy is unfolding in Zambia, which is about to lose ZESCO, Zambia's predominant power supply company that generates 80% of Zambia's electricity consumed in that country, and a strategic country asset, as a loan default payment to a China lender (Laterza & Mususa, 2018).

The SDGs, the New Urban Agenda and the Urban Economy

The United Nations has been in the forefront in supporting actions and strategies aimed at sustainability, equity, resilience, social inclusion and economic growth. This is amplified in its blueprint for the 17 *Sustainable Development Goals* (*SDG*s) that captures and addresses the majority of the global challenges that impact upon our Earth (United Nations, 2018). To sustain the livelihood of the world's communities, the target for the achievement of the goals is 2030. Relevant to cities is *SDG 11* that is geared towards sustainable cities and communities that are aspirationally to be 'inclusive, safe, resilient and sustainable' (UN, 2015). Country responses tend to address rapid urbanization patterns and mass migration movements to cities, which, the UN acknowledges will rise in numerical population to approximately 5 billion by 2030 (UN-Habitats, 2016). With such a boom in demography, a myriad of challenges is expected.

Position unprecedented populated growth with unstable climate change effects and you reach a state that former US Vice President Al Gore has labelled in 2006 an '*Inconvenient Truth*' (Gore, 2006). While *SDG 11* was formulated in anticipation of 2030, the majority of the urban dwellers are already experiencing massive waste concentrations because the world is experiencing an increasing consumption patterns leading to the generation of large amounts of waste (Allam, 2017a, 2018c; Allam & Jones, 2018b; Hoornweq & Bhada-Tata, 2012). There is also an increasing notable impetus to counter the associated risks of shortages in potable water supplies, of food security and of basic social services, especially in slums and informal settlements that are somehow always neglected in policy documents (Mahabir, Crooks, Croitoru, & Agouris, 2016). There is also the anticipation that vehicular transportation use will continue to rise and pose higher traffic congestion issues and pollution emission discharges at much higher magnitudes when placed in context with human health-related risks (Wang, Mao, Li, Xiong, & Wang, 2015). This is without even considering our animal and vegetation stakeholders that are being equally negatively affected by these impacts. In addition, the provision for decent and affordable housing has been a challenge, and it is anticipated to continue as more people move to urban centres as documented by UN-Habitats (2015b), and by Van Noorloos and Kloosterboer (2018). Therefore, *SDG 11* is meant to channel the attention of governments, city managers and other human stakeholders to strategize and design policies that ensure decent and affordable human housing in conjunction with providing clean, safe and greener environments and urban landscapes. Echoing this, *SDG 11* also supports the need for the adoption of alternative and safe energy sources that can better lead to the reduction of greenhouse gases (GHG) and other forms of pollutions. All these objectives point to making cities more resilient against the evident impacts of climate change while scaffolding them as more liveable environments.

To ensure the actualization and achievement of the *SDG*s in cities, the *New Urban Agenda* (*NUA*) was proposed (United Nations, 2016). The *NUA* was adopted at the United Nations *Conference on Housing and Sustainable Urban Development* (*Habitat III*) in Quito, Ecuador, on 20 October 2016; a year after the adoption of the *SDG*s in Paris at the United Nation's 2030 *Agenda for Sustainable Development Summit* (United Nations, 2018). The *NUA* dwells on the subject of how cities, towns and rural areas are to be planned and managed so as to achieve *SDG*s and the *Paris Agreement* on climate change (United Nations, 2015). The *NUA* is anchored in five main pillars that focus upon ensuring quality planning, construction, development, management and the general improvement of urban fabric. The five pillars include national urban policies, urban legislation and regulations, urban planning and design, local economy and municipal finance and local implementation (United Nations, 2016). These pillars underscore measures that every country should adopt to cooperatively enabling urbanization rather than succumbing to it. Dugarova and Gülasan (2017) argue that the *NUA* offers a roadmap for cities to follow as they progressively focus upon improving their prosperity through innovation and economic growth, and in servicing their roles as centres of human culture and social well-being. The success of the *NUA*, as highlighted by UN-Habitats (2016), will depend upon the quality of financial frameworks that cities in different economies adopt, as well as how they manage to command cooperation from all human stakeholders in the development agenda. By implementing these strategies, urbanization in cities is bound to bring about job creation, improvements to the liveability of cities by ensuring better human quality of life and improvements to the economic situations of cities, countries and regions.

The *NUA* observes that if all cities propose to attain equity by re-addressing their planning, designing, financing and management activities amongst other things, they will be venturing into improving our human settlements resulting in potential shifts towards negating and mediating poverty and hunger-free cities. Such measures could reduce inequalities and actualize *SDG 11* and other goals, like those goals that focus upon women and girls, and seek to protect the environment, amongst others.

True to the commitment made in the *NUA*, progressive results in all aspects of cities and human lives have been attained as highlighted in the *Sustainable Development Goals Report* of 2018 (United Nations, 2018). The report reveals that 152 countries have readdressed their national urban policies towards engaging in sustainable urbanization initiatives and processes as prescribed in the *NUA*. This includes statistics that over 204 municipalities in 103 countries have managed to implement waste management strategies and are now having comprehensive collection, sorting and management strategies in place. The report also evidences that there have been steady investments in drainage infrastructure in flood-prone areas, and in regulations about re-dressing unsustainable land-use practices. The latter two aspects are key to the survival of SIDS, and to protecting their vital infrastructures that are key to their economic stability.

In contrast, it is also reported that the number of slum dwellers is still increasing and the majority of these slums operate on and subsume formal and or informal public open spaces that are deterring development that is a catalyst to the growth of these urban economies (UN-Habitats, 2016). Similarly, in many countries air pollution quality has not reduced and is still 2.5 times higher than the safe threshold (Kelly & Fussell, 2015; United Nations, 2018). These negative indicators have been reported to majorly affect low-income economies and SIDS, and further point to the need for targeted investment in different urban sectors. The data also demonstrate at why mitigation infrastructure and strategies need to be re-aligned to ensure adherence with both *SDG*s and the *NUA*.

Though the *SDG* goals and *NUA* targets set in these documents are bound to transform the development landscape of cities, their implementation depends upon the financing models that each country adopts (Africa Development Bank, 2018). To developed countries and those with larger GDPs, it is a matter of implementation. On the other hand, for most SIDS and low-income economies, the financing of these goals is a daunting task and finding a trustworthy loan partner is equally a challenge. According to the UNCTAD (2018), most of the available financing models are tied to stringent terms and conditions that most of these economies are unable to cope with. Similarly, some are 'debt traps', and some countries have already been snared (Watson & Kellett, 2016).

Harris and Lane (2018) perceive them as obstacles to the achievement of SDGs.

Foreign Aid and the Debt Trap Diplomacy

Gurara et al. (2017) underline that public spending in infrastructural development in most low-income economies and SIDS has not been sufficient to address priorities. This is despite the recognition that such spending in the public realm is the most potent financing model for such projects. To complement this, most of these countries have turned to internal funding where they have tried public–private partnership (PPP) models that have only succeeded in a limited number of cases. A notable feature of such PPP projects is the magnitude of required investment that has to be calibrated to match the prescriptions of the *SDG*s and the *NUA*. In addition to the cost and size magnitude of projects requiring financing beyond what private partners can manage, some governments host negative credits when applied to their national policies relating to the adoption of economic growth strategies (Estache, Serebrisky, & Wren-Lewis, 2015). Those low-income economies and SIDS that have benefited from internal financing are characterized by strong and stable political environments and leadership that promote economic resilience, like in case of Singapore (Ministry of Foreign Affairs Singapore, 2018), or in the case of the remote African island country of Mauritius (Allam & Newman, 2018a).

However, for most economies, in these categories, external financing is deemed as the only and most appropriate option since there are numerous financial institutions that have customizable infrastructural financing models. The Africa Development Bank (2018) highlights that institutional investors and commercial banks globally have various mechanisms on offer for financial support, spurring over USD$100 trillion in loans. The advantage with external financing, over other models, is that countries have the ability to secure long-term loans that allow them to invest in high-quality infrastructure at low costs (Farquharson et al., 2011). Similarly, most financiers are from rich and developed countries that have already finances but also an excess number of professionals, experts and consultants that can intellectually and technically benefit borrowers.

Despite having these financiers ready to invest in development projects, low-income countries and SIDS are still witnessing a low influx of capital. Tyson (2018) attributes this to a number of factors including the lack of 'bankable projects', unstable political and macro-economic environments that characterize most of these economies, and the lack of or mismatch between what is offered as collateral to satisfy the loan conditional requirements of institutional investors. Bationo, Griffith-Jones, Murinde, Soumaré, and Tyson (2018) highlight that the small Gross Domestic Products (GDPs), that characterize most low-income economies and some SIDS economies, undermine their business fundamentals. Thus, a majority of financers are opting to focus capital investments elsewhere (Carter & Tyson, 2015). They also attribute the low financing to the inherent risks posed by low economic growth, political instabilities and other anti-business practices that most developing economies have been historically and culturally beholden to. Flynn, Rao, Horner, and Gashi (2018) express that a majority of economies experience many risks, like political, regulatory, macro-economic and technical risks, that discourages financiers, or force them to introduce stringent terms and conditions. A more serious and notable issue is the perceived un-creditworthy of many economies (Allam & Jones, 2019; OECD, 2014b).

With all these challenges and barriers to securing loans from private institutions, many low-income and SIDS economies are turning towards foreign aid and loans from individual countries like China, of which China seems to overlook some of these aforementioned risks (Were, 2018). Unlike other financiers, this form of loan attracts higher interest rates, are attached to the lenders' capital interests, and are more likely to be defaulted (Senators, 2018; Were, 2018). In respect to loans from China, Michael et al. (2016) admit that such loan arrangements often lack institutionalized and established support systems like traditional loan arrangements and are usually concessional, but at lower rates. Their attractiveness and ease of securement have put many low-income countries at risk of defaulting since most of these arrangements are substantively and legally obligatory upon these countries than their actual economies can support. On an unfortunate note, most of these loan arrangements are tied to strategic assets or natural resources that most country's risk handing over once the burden of repayment gets heavier (Crane, Albrecht, Duffin, & Albecht, 2018; Payer, 1975).

Repayment burdens that low-income countries are facing, and the risk of losing some of their vital assets to Chinese owned companies, are amplified in a letter addressed to United States' Secretary of Treasury and Department of State authored by Mnuchin and Pompeo (Senators, 2018). This letter reveals that a number of low-income countries, that have to date benefited from Chinese loan arrangements, are on the verge of losing vital infrastructure and were entering into discussions with the International Monetary Fund (IMF) seeking bailout monies to prevent them from defaulting. An example is Sri Lanka, which was bailed out in 2016 from a bad debt of USD$1.5 billion, and in 2017, failed again resulting in the handing over of the Hambantota Port to China for a period of 99 years (Abi-Habib, 2018). Additionally, the country of Djibouti is reported to be on the brink of also handing over its Doraleh Container Terminal, in the Port of Doraleh, to China's Export-Import Bank (EXIM Bank) (Senators, 2018), and Pakistan is also reported to be at risk of handing to Chinese loan brokers its Arabian Seaport at Gwadar due to its over-reliance upon Chinese loans (Aamir, 2017).

Such challenges of debt burden, and the seizure of vital assets, are surmountable if SIDS and low-income economies address some of the pressing risks of defaulting including enforcing private financiers to use structured, institutionalized and development-oriented financing models (Michael et al., 2016).

One area that needs to be addressed in these arrangements is the political sphere, especially where the political model of regime changes in most of the economies is fragility and tenuous (Carter & Tyson, 2015). Thus, it has been observed that countries with uncertain or short-lived political regimes, typically most low-income countries and some SIDS, have unviable and stable long-term economic planning strategies because every time a regime changes each new regime comes in with new plans and agendas, resulting in most existing projects being left in limbo because they are tainted with messages of the previous regime(s) (Islam, 2005). Such is particularly evident because most of the economic activities and practices of these countries are controlled by (military and or economic) discrete political classes. The decision to borrow and at what terms and from whom is thus majorly influenced by local politics which in itself is guided by a range of geopolitical factors.

Introducing Fiscal Incentivization and Development

The burden of debt that a majority of low-income economies and SIDS face from external lenders is avoidable. Where they exist, it is often insurmountable. While a myriad of strategies can be adopted to finance infrastructure development projects, either externally or internally, the utilization of available fiscal tools like tax incentives and fiscal packages are amongst the most rewarding (Begashaw & Shah, 2017). This is because these tools allow governments to catalyse development through other means even though they may not be able to raise capital, through direct and traditional means. As demonstrated from the literature above, and from a wide array of discussions and reports on this topic (Africa Development Bank, 2018; Bationo et al., 2018; Carter & Tyson, 2015; PWC, 2016), it has been established that most low-income countries and SIDS are unable to attract substantial infrastructure development financing. This is due to a wide range of issues including corruption, different forms of risks, governance issues and weak institutions amongst others.

Flynn et al. (2018) point out that those entities willing to offer any form of financial support to low-income economies and SIDS do so with considerable stringent measures and terms. Although this is the typical case, the Africa Development Bank (2018) outlook report underscores that most of these economies offer lucrative investment opportunities that investors can now capitalize upon. But, through the ADB Avenue, these investors require some assurances and guarantees of tangible returns and a favourable tax mechanism. These guarantees are warranted by the difficulties that low-income economies and SIDS contend with due to their small GDPs, their political and business environments, and their poor credit scores, amongst others.

The introduction of fiscal incentives and packages would be a safe starting point for low-income economies and those SIDS struggling with raising enough capital for investments aimed towards sustainable development and addressing climate change. These incentives could be in the form of tax exemptions, lower tax rates and tax concessions amongst others, as well as inclusion within special economic zones (SEZs), as highlighted by Sinenko (2016). A PWC (2016) report, that focused upon Malawi, pointed to the importance of fiscal incentives by highlighting their importance in compensating for constraints and deficiencies such as costs, delays, risks and low ranking in formulating business indexes.

Offering fiscal incentives and packages in the case of infrastructural investments also has the potential to increase financial competition amongst lenders, all at the benefit of the host country (Klemm & Stefan, 2009). PWC (2016) underscores that this competitive environment is important because it influences the host country's performance both domestically and in the international arena.

These fiscal tools have been effective in SEZ and ports due to the unique contribution of these areas in a country's economic growth in terms of budgetary revenues and employment. The case of Mauritius on this front is commendable (Allam, 2017b, 2019a, 2019b; Allam & Newman, 2018a; Dabeedooal, Dindoyal, Allam, & Jones, 2019). According to PWC (2016), tax incentives are not always effective in a country's investment projections due to issues like structuring deficiencies, monetary abuses and sometimes alternate investor-oriented options. Thus, they are majorly concentrated in SEZs and ports that have a higher potential to yield positive results. These SEZs have high a potential to attract Foreign Direct Investments (FDIs) (Chakraborty, Kathuria, & Gundimeda, 2017; Crane et al., 2018). SEZs, especially when integrated into ports, are easier for governments to formulate incentives geared towards optimization of all economic benefits that they can deliver. Incentives can also be offered in these SEZs to aid and stimulate greener practices that are in line with *SDG*s, especially given that most of these SEZs involve high levels of consumption of resources including energy power (Adani, 2017). With fiscal incentives, as explained by Sinenko (2016), SEZs are able to increase outputs leading to growth in and involvement by both local and foreign companies and corporations. This growth in turn can stimulate increased income as companies and corporations are able to make attractive profits due to reductions in tax burdens (Shah, 1995; Sinenko, 2016; Zeng, 2015). The backdrop of these outcomes is the potential growth in in-country employment opportunities, which most countries will benefit from, where SEZs are located, and which can stimulate local economies directly or indirectly (Engman, Onodera, & Pinali, 2007).

However, the use of fiscal tools in SEZs has been discredited in some quarters as being counterproductive. These criticisms are predicated upon arguments that they are perceived as denying governments the opportunity to earn revenue which investors could have paid if they were outside SEZs (Bationo et al., 2018). Additionally, while SEZs have a decades-long history around the world most researchers conclude that their success is due to a variety of reasons.

While these criticisms are partly true, in the case of infrastructure development financing in SIDS and low-income economies, capital in the form of loans has often been reported in the media as being embezzled or utilized in alternate different ways more often not connected to the actual project. With fiscal incentives in place, these government entities do not earn much tangibly in revenue, but do earn major tangible benefits in high-quality infrastructure being constructed at lower costs than they could achieve themselves. The benefits from this strategy include the creation of local employment and on-site training opportunities in large infrastructural projects (Sinenko, 2016; Tyson, 2018). Therefore, forfeited present revenue translates to future economic and societal gains. With Foreign Direct Investments (FDIs) and PPPs backed by fiscal incentives, the obvious shortcomings including weak public institutions, waste of resources, poor governance and low implementation that can be easily overcome because the private sector is self-regulatory capacity (Daseking & Kozack, 2003). Laboul and Croce (2014) explain that using fiscal tools to attract investors in infrastructure investment succeeds because it allows private investors a sense of assurance regarding risk appraisals while ensuring that public benefits from vital infrastructure that can be geared towards improving societal needs.

This model, of using fiscal tools to influence results such as creation of employment and the increase of national income, is an ideal strategy that can be adopted by SIDS and low-income countries to finance their infrastructure development projects. By offering incentivization, SIDS and low-income country governments could not only attract foreign investors, but also encourage public–private partnerships (PPP); the latter is increasingly viewed by researchers as one of the most successful approaches towards the achieving *SDG*s.

This strategy, of fiscal incentivization, has been adopted in Mauritius, a SIDS economy, and has evidenced positive economic growth in a short time scale. This has been documented by (M. Z. Allam, 2018; Allam & Newman, 2018a, 2018b). What is important in this context is how to further encourage these types of investments, both local and foreign, towards projects and objectives as defined by governments.

A common misconception appears to lie in the categorization of infrastructure development works that mischievously means solely 'hard infrastructure'. With the advent of technology, numerous digital solutions and soft interventions can be encouraged with substantial impacts on savings in urban management (Allam, 2014, 2017b). This recommendation is tabled without discussing the merits of green and blue-green infrastructure initiatives that can be sustainable rich but not involve extensive 'hard infrastructure'. In this regard, fiscal mechanisms could be calibrated for urban areas requiring significant investment for regenerative purposes, while at the same time supporting sustainable and resilient processes and community needs.

Conclusion

This paper explored an alternative model to achieve sustainability and resilience in low-income economies and SIDS that usually face the challenges of climate change but are unable to respond to them without increasing debt. Fiscal mechanisms are suggested to be encouraged, like through Special Economic Zones (SEZs), but on a larger urban and regional scale. These fiscal mechanisms will ensure private sector participation towards a locally defined public vision, as the former will be particularly interested in pursuing economic gains from fiscal incentives. While we consider physical infrastructure development, or establishing a Special Economic Zone (SEZ), more superior to relying on internal public–private partnerships (PPPs) or foreign loans from abroad, it is very evident that in SIDSs economies that medium to large-scale infrastructure can be instrumental. This conclusion is demonstrated in the role SEZs have played in the Mauritius and Singapore economies. Further, providing an environment of trust and political stability, and increased innovative PPP collaboration opportunities, can serve as supporting mechanisms to aid sustainability and SDGs while still recognizing that are highly dependent upon government backing and their loan capacity despite loans being tailored to cater for both local visions and private economic needs. Foreign aid, loans and traditional funding mechanisms are shown as being unresponsive to current contexts and unsustainable in a broader long-term vision. As such, there is a need to think of alternative ways to fund development works in countries that can least afford them, and while doing so ensuring that they respond to sustainability measures as supported by both the *SDG* 11 and the *New Urban Agenda*.

References

Aamir, A. (2017, December 27). *Will Gwadar go the way of Hambantota? Why Chinese loans to Pakistan are sparking takeover fears along the economic corridor*. Retrieved from https://www.scmp.com/comment/insight-opinion/article/2125783/will-gwadar-go-way-hambantota-why-chinese-loans-pakistan-are.

Abi-Habib, M. (2018, June 25). How China got Sri Lanka to cough up a port. *New York Times*.

Adani. (2017). *Sustainability Report 2016–17*.

Africa Development Bank. (2018). *African economic outlook 2018*.

Aliyu, A. A., & Amadu, L. (2017). Urbanisation, cities and health: The challenges to Nigeria—Review. *Annals of African Medicine, 16*(4), 149–158.

Allam, M. Z. (2018). *Redefining the smart city: Culture, metabolism and governance*. Case Study of Port Louis, Mauritius (PhD). Curtin University, Perth, Australia. Retrieved from https://espace.curtin.edu.au/handle/20.500.11937/70707.

Allam, Z. (2012). Sustainable architecture: Utopia or feasible reality? *Journal of Biourbanism, 2*, 47–61.

Allam, Z. (2014). Exploring the urban heat island (UHI) effect in Port Louis, Mauritius. *University of Mauritius Research Journal, 20*, 138–153.

Allam, Z. (2017a). Building a conceptual framework for smarting an existing city in Mauritius: The case of Port Louis. *Journal of Biourbanism, 4*(1–2), 103–121.

Allam, Z. (2017b). *A theoretical application of the Extended Metabolism Model in Port Louis in a bid to promote urban sustainability*. Paper presented at the 2nd International Conference on Energy, Environment and Climate Change (ICEECC 2017), Mauritius.

Allam, Z. (2018a). Contextualising the smart city for sustainability and inclusivity. *New Design Ideas, 2*(2), 124–127.

Allam, Z. (2018b). On smart contracts and organisational performance: A review of smart contracts through the blockchain technology. *Review of Economic and Business Studies, 11*(2), 137–156.

Allam, Z. (2018c). The zero-waste city: Case study of Port Louis, Mauritius. *International Journal of Sustainable Building Technology and Urban Development, 9*(3), 110–123.

Allam, Z. (2019a). Enhancing renewable energy adoption in megacities through energy diversification, land fragmentation and fiscal mechanisms. *Sustainable Cities and Society, 1*, 101841.

Allam, Z. (2019b). Identified priorities for smart urban regeneration: Focus group findings from the city of Port Louis, Mauritius. *Journal of Urban Regeneration & Renewal, 12*(4), 376–389.

Allam, Z., & Dhunny, Z. A. (2019). On big data, artificial intelligence and smart cities. *Cities, 89*, 80–91.

Allam, Z., & Jones, D. (2018a). *Promoting resilience, liveability and sustainability through landscape architectural design: A conceptual framework for Port Louis, Mauritius: A small island developing state*. Paper presented at the IFLA World Congress Singapore 2018, Singapore.

Allam, Z., & Jones, D. (2018b). Towards a circular economy: A case study of waste conversion into housing units in Cotonou, Benin. *Urban Science, 2*(4), 118.

Allam, Z., & Jones, D. (2019). Climate change and economic resilience through urban and cultural heritage: The case of emerging small island developing states economies. *Economies, 7*(2), 62.

Allam, Z., & Newman, P. (2018a). Economically incentivising smart urban regeneration: Case study of Port Louis. *Mauritius. Smart Cities, 1*(1), 53–74.

Allam, Z., & Newman, P. (2018b). Redefining the smart city: Culture, metabolism & governance. *Smart Cities, 1*, 4–25.

Arafat, W., Bing, Z. Y., & Al-Mutawakel, O. (2018). Infrastructure developing and economic growth in United Arab Emirates. *Business and Economic Research, 8*(1), 95–114.

Asian Development Bank. (2014). *A comparative infrastructure development assessment of the Republic of Korea and the Kingdom of Thailand*. Mandaluyong City, Philippines: Asian Development Bank.

Attapattu, A. M. C. P., & Padmasiri, H. M. N. (2018). Long run effect of public debt on economic growth in Sri Lanka. *Modern Economy, 9*(4), 775–789.

Bah, E. M., Faye, I., & Geh, Z. F. (2018). Slum upgrading and housing alternatives for the poor. In E. M. Bah, I. Faye, & Z. F. Geh (Eds.), *Housing market dynamics in Africa*. London: Palgrave Macmillan.

Bakar, N. A. A., & Mat, S. H. C. (2017). Infrastructure development and economic growth in Malaysia: An ARDL approach. *International Journal of Multidisciplinary Research and Development, 4*(10), 51–55.

Bationo, F., Griffith-Jones, S., Murinde, V., Soumaré, I., & Tyson, J. (2018). *Capital flows and productivity in Africa: The angel is in the details*. London: SOAS, University of London.

Begashaw, B., & Shah, A. (2017). *SDG financing for Africa: Key propositions and areas of engagement*. Paper presented at the The Sustainable Development Goals Center for Africa Conference, Kigali, Rwanda.

Burton, P., Lyons, K., Carol Richards, Amati, M., Rose, N., Fours, L. D., ... Barclay, R. (2013). *Urban food security, urban resilience and climate change*. Gold Coast.

Capdevila, I., & Zarlenga, M. I. (2015). Smart city or smart citizens? The Barcelona case. *Journal of Strategy and Management, 8*(3), 266–282.

Carter, P., & Tyson, J. (2015). *Post-monterrey financing trends in sub-Saharan Africa*. Addis Ababa: United Nations Economic Commission for Africa (UNECA).

Chakraborty, T., Kathuria, V., & Gundimeda, H. (2017). Have the special economic zones succeeded in attracting FDI? *Analysis for India. Theoritical Economics Letters, 7*(3), 623–642.

Clements-Croome, D., Marson, M., Yang, T., & Airaksinen, M. (2017). Planning and design scenarios for liveable cities. In M. A. Abraham (Ed.), *Encyclopedia of sustainable technologies* (pp. 81–97). Amsterdam: Elsevier.

Crane, B., Albrecht, C., Duffin, K. M., & Albecht, C. (2018). China's special economic zones: An analysis of policy to reduce regional disparities. *Regional Studies, Regional Science, 5*(1), 98–107.

Dabeedooal, J. Y., Dindoyal, V., Allam, Z., & Jones, S. D. (2019). Smart tourism as a pillar for sustainable urban development: An alternate smart city strategy from Mauritius. *Smart Cities, 2*(2), 153–162.

Daseking, C., & Kozack, J. (2003). Avoiding another debt trap: Low-income countries need a sustainable borrowing strategy if they are to achieve development goals. *Finance & Development*, 20–23.

Dong, G., Weng, B., Qin, T., Yan, D., Wang, H., Gong, B., … Xing, Z. (2018). The impact of the construction of sponge cities on the surface runoff in Watersheds, China. *Advances in Meteorology, 9*, 1–9.

Dugarova, E., & Gülasan, N. (2017). *Global trends: Challenges and opportunities in the implementation of the sustainable development goals*. New York, NY.

Ehlers, T. (2014). *Understanding the challenges for infrastructure finance* (BIS Working Papers 454, 454). Bank for International Settlements.

Eko Atlantic. n.d. *An investment opportunity on an unprecedented scale*. Retrieved from https://www.ekoatlantic.com/about-us/.

Emilsson, T., & Sang, O. A. (2017). *Impacts of climate change on urban areas and nature-based solutions for adaptation*. Cham: Springer.

Engman, M., Onodera, O., & Pinali, E. (2007). *Export processing zones: Past and future role in trade and development* (OECD Trade Policy Working Papers No. 53).

Estache, A., Serebrisky, T., & Wren-Lewis, L. (2015). Financing infrastructure in developing countries. *Oxford Review of Economic Policy, 31*(3–4), 279–304.

Farquharson, E., Mästle, C. T. D., Yescombe, E. R., & Encinas, J. (2011). *How to engage with the private sector in public-private partnerships in emerging markets*. Washington, DC: The International Bank for Reconstruction and Development/The World Bank.

Flynn, M., Rao, A. K., Horner, J., & Gashi, D. S. (2018). Smart cities funding and financing in developing economies: Assisting development cities to finance their infrastructure gap through private sector participation apporaches.

Gore, A. (2006). *An inconvenient truth: The planetary emergency of global warming and what we can do about it.* New York: Rodale.

Gurara, D., Klyuev, V., Mwase, N., Presbitero, A., Xu, X. C., & Bannister, G. (2017). *Trends and challenges in infrastructure investment in low-income developing countries.* Washington, DC: International Monetary Fund.

Hales, S., Baker, M., Howden-Chapman, P., Menne, B., Woodruff, R., & Woodward, A. (2007). Implications of global climate change for housing, human settlements and public health. *Review on Environmental Health, 22*(4), 295–302.

Harris, E., & Lane, C. (2018, April 27). *Debt as an obstacle to the sustainable development goals.* Retrieved from https://blogs.imf.org/2018/04/27/debt-as-an-obstacle-to-the-sustainable-development-goals/?cid=sm-com-TW.

Hoornweq, D., & Bhada-Tata, P. (2012). *What a waste: A global review of solid waste management.* Washington, DC: World Bank.

Hughes, S., & Sarzynski, A. (2015). Building capacity for climate change adaptation in urban areas. *Urban Climate, 13,* 1–3.

Islam, M. N. (2005). Regime changes, economic policies and the effect of aid on growth. *The Journal of Development Studies, 41*(8), 1467–1492.

Kelly, F. J., & Fussell, J. C. (2015). Air pollution and public health: Emerging hazards and improved understanding of risk. *Environmental Geochemistry and Health, 37*(4), 631–649.

Klemm, A., & Stefan, V. P. (2009). *Empirical evidence on the effects of tax incentives.* Washington, DC: Brookings Institution.

Kodongo, O., & Ojah, K. (2016). Does infrastructure really explain economic growth in sub-Saharan Africa? *Review of Development Finance, 6*(2), 105–125.

Laboul, A., & Croce, R. D. (2014). *Private financing and government support to promote long-term investments in infrastructure.* Paris: OECD.

Lafortezza, R., Chen, J., Van den Bosch, C. K., & Randrup, T. B. (2018). Nature-based solutions for resilient landscapes and cities. *Environmental Research, 165,* 431–441.

Laterza, V., & Mususa, P. (2018, October 11). *Is China really to blame for Zambia's debt problems?* Retrieved from https://www.aljazeera.com/indepth/opinion/china-blame-zambia-debt-problems-181009140625090.html.

Lee, D. (2014). *Africa is ready to leapfrog the competition through smart cities technologies.* Johannesburg: Deloitte.

Lim, C., Kim, K.-J., & Maglio, P. P. (2018). Smart cities with big data: Reference models, challenges, and considerations. *Cities, 82,* 86–99.

Mahabir, R., Crooks, A., Croitoru, A., & Agouris, P. (2016). The study of slums as social and physical constructs: Challenges and emerging research opportunities. *Regional Studies, Regional Science, 3*(1), 737–757.

McDonald, C. (2017). *Current and future impacts of climate change on housing, buildings and infrastructure.* Surry Hills, NSW.

Michael, E., Omuruyi, M., Zhexi, W., Bing, Z. Y., Kai, W., & Xin, X. (2016). China's infrastructure development and its impact on Africa economic growth. *International Journal of African and Asian Studies, 23,* 36–47.

Miller, J. D., & Hutchins, M. (2017). The impacts of urbanisation and climate change on urban flooding and urban water quality: A review of the evidence concerning the United Kingdom. *Journal of Hydrology: Regional Studies, 12,* 345–362.

Ministry of Foreign Affairs Singapore. (2018). *Towards a sustainable and relient Singapore.* Retrieved from https://sustainabledevelopment.un.org/content/documents/19271Singapores_Voluntary_National_Review_Report.pdf.

Nadali, P., Thomas, R. M., & Taleb, H. (2015). Rethinking and redesigning sustainability in Dubai's sustainable city. *WIT Transactions on Ecology and the Environment, 193,* 997–1005.

OECD. (2014a). *Cities and climate change: National governments enabling local action.* OECD Policy Perspectives.

OECD. (2014b). *Pooling of institutional investors capital-selected case studies in unlisted equity infrastructure.* OECD, 1–64. Retrieved from https://www.oecd.org/finance/OECD-Pooling-Institutional-Investors-Capital-Unlisted-Equity-Infrastructure.pdf.

Palanivel, T. (2017, September 6). *Rapid urbanisation: Opportunities and challenges to improve the well-being of societies* (Human Development Reports). Retrieved from http://hdr.undp.org/en/content/rapid-urbanisation-opportunities-and-challenges-improve-well-being-societies.

Palei, T. (2015). Assessing the impact of infrastructure on economic growth and global competitiveness. *Procedia Economics and Finance, 23,* 168–175.

Parker, J., & Simpson, G. D. (2018). Public green infrastructure contributes to city livability: A systematic quantitative review. *Lands, 7,* 161–184.

Payer, C. (1975). *The debt trap: The international monetary fund and the third world.* New York: New York University Press.

Pioletti, M. (2016). Cities and development. *Urban Research & Practice, 9*(3), 329–331.

PWC. (2016). *A comprehensive analysis and review of investment, production and export incentives in Malawi.* Blantyre, Malawi: PWC.

Schmidt-Traub, G., & Sachs, D. D. (2015). *Financing sustainable development: Implementing the SDGs through effective investment strategies and partnerships.*

Senators. (2018). *US Senate China IMF Letter.* Washington, DC.

Sepasgozar, S. M. E., Hawken, S., Sargolzaei, S., & Foroozanfa, M. (2018). Implementing citizen centric technology in developing smart cities: A model for predicting the acceptance of urban technologies. *Technological Forecasting and Social Change, 142,* 105–116.

Shah, A. (1995). *Fiscal incentives for investment and innovation.* New York: Oxford University Press.

Shami, M., & Majid, H. (2014). *The political economy of public goods provision in slums*. London: International Growth Centre (IGC).

Shi, Y., Guo, S., & Sun, P. (2017). The role of infrastructure in China's regional economic growth. *Journal of Asian Economics, 49,* 26–41.

Siew, G., & Allam, Z. (2017). *Culture as a driver for sustainable urban development*. Paper presented at the UIA 2017 Seoul World Architects Congress, Seoul. http://www.uia2017seoul.org/P/papers/Abstract/Paper/Oral/PS2-52/O-0091.pdf.

Sinenko, O. A. (2016). Methods of assessing of tax incentives effectiveness in special economic zones: An analytical overview. *Journal of Tax Reform, 2*(3), 168–178.

Smit, S., Musango, J. K., Kovacic, Z., & Brent, A. C. (2017). Conceptualising slum in an urban African context. *Cities, 62,* 107–119.

Srinivasu, B., & Rao, P. S. (2013). Infrastructure development and economic growth: Prospects and perspectives. *Journal of Business Management & Social Sciences Research, 2*(1), 81–91.

Steadman, I. (2013, January 25). Kenya breaks ground on its 'Silicon Savannah' city project. *Technology*. Retrieved from https://www.wired.co.uk/article/kenya-silicon-savvanah.

Tyson, J. E. (2018). *Private infrastructure financing in developing countries* (Working Paper No. 553). ODI.

UN. (2015). Goal 11: Make cities inclusive, safe, resilient and sustainable. *Sustainable Development Goals*. Retrieved from https://www.un.org/sustainabledevelopment/cities/.

UNCTAD. (2017). *The Least Developed Countries Report 2017*. New York and Geneva: United Nations.

UNCTAD. (2018). *Scaling up finance for the sustainable development goals: Experimenting with models of multilateral development banking*. New York and Geneva: UNCTAD.

UNDP. (2016). *Sustainable urbanization strategy: UNDP's support to sustainable, inclusive and resilient cites in the developing world*. New York, NY: United Nations Development Programme.

UN-Habitats. (2015a). *E-Governance and urban policy design*. Retrieved from http://unhabitat.org/books/e-governance-and-urban-policydesign-in-developing-countries/.

UN-Habitats. (2015b). *Urbanization and climate change in small island developing states*. Nairobi, Kenya. Retrieved from https://unhabitat.org/wpdm-package/urbanization-and-climate-change-in-small-island-developing-states/?wpdmdl=114762.

UN-Habitats. (2016). *Urbanization and development: Emerging futures*. Nairobi: UN-Habitats.

United Nations. (2015). Paris Agreement. *UNFCCC.* Retrieved from https://unfccc.int/sites/default/files/english_paris_agreement.pdf.

United Nations. (2016). *The new urban agenda.* Paper presented at the United Nations Conference on Housing and Sustainable Urban Development (Habitat III), Quito, Ecuador.

United Nations. (2018). *Sustainable Development Goals Report.* New York.

Van Noorloos, F., & Kloosterboer, M. (2018). Africa's new cities: The contested future of urbanisation. *Urban Studies, 55*(6), 1223–1241.

Wang, J., Mao, Y., Li, J., Xiong, Z., & Wang, W.-X. (2015). Predictability of road traffic and congestion in urban areas. *PLoS ONE, 10*(4), e0121825.

Watson, C., & Kellett, J. (2016). *Finacing sustainable development: The critical role of risk and resilience.*

Weerakoon, D. (2017). Sri Lanka's debt troubles in the new development finance landscape. *Third World Thematics: a TWQ Journal, 2*(6), 744–761.

Were, A. (2018). *Debt traps? Chinese loans and Africa's development options* (Policy Insights 66). SAIIA.

World Bank. (2010, December). *Cities and climate change: An urgent agenda* (Urban Development Series Knowledge Papers, 63704, 10). Washington, DC.

World Bank Group. (2014). *Sovereign wealth funds and long-term development finance: Risks and opportunities.* Paper for the G20 Investment and Infrastructure Working Group, World Bank.

Zaidan, E., & Kovacs, J. F. (2017). Resident attitudes towards tourists and tourism growth: A case study from the Middle East, Dubai in United Arabs Emirates. *European Journal of Sustainable Development, 6*(1), 291–307.

Zeng, D. Z. (2015). *Global experiences with special economic zones* (World Bank Policy Research Working Paper 7240).

CHAPTER 5

Achieving Urban Resilience Within the Capitalist Movement

Abstract This concluding chapter dwells into the theories and content that underpin this book. As a consequence, it explores, as an essay, how thematically to achieve resilience and sustainability without disrupting current economic models. Recognizing that no disruptions to local, regional and global economies will encourage a wider adoption of the proposed model, the consequential sustainability outcomes can still be wider in their encouragement of a more sustainable transition within the current capitalist culture. The final chapter thus explores how the proposed model can be applied to navigate through the contemporary capitalist culture, and how this culture can benefit from this approach as opposed to a disruptive shift that could destabilize and affect global markets.

Keywords Resilience · Sustainability · Economics · Marxism · Capitalist · Sustainable transition

Introduction

At the height of Industrial Revolution, Karl Marx warned that the world was headed into an era where the capitalist movement would thrive leading to inequalities and societal problems (Marx & Engels, 2018). True to

Z. Allam et al., *Cities and Climate Change*,
Palgrave Studies in Climate Resilient Societies,
https://doi.org/10.1007/978-3-030-40727-8_5

this, two centuries later, the world is controlled by capitalistic economies. In his arguments, he narrated observations about the souring of relationships between capitalists and the proletariat. Even in this modern era, there is little change in how this tension is viewed, and it has been expanded even further to remote economies external to Europe that was the study venue for Marx. From a survey of literature, a wide range of sources (Ali, 2016; Patomäki, 2017) support the conclusion that there are economies; especially those in the Global North that have assumed capitalistic characteristics and use their influence to grow their economies, at the same time as gaining political mileage and control. As these agendas have and continue to be pursued, their actions, such as the use of non-renewable energy and excessive extraction of resources from the ecosystem, are leading to obvious negative impacts compromising the world's climate and its environment resulting in the current global economic crisis that is hurting the world. While this agenda furthers, their counterparts in the Global South, either developing or least developing economies, continue to reel in poverty, face poor economic growth and bear the impacts of climate change (Allam & Jones, 2019; Kling, Lo, Murinde, & Volz, 2018; Tol, 2018; Wright, Reeves, & Huq, 2016). This is despite concerted global efforts, like the *Paris Agreement*, to address some of the above challenges.

The expected changes that could introduce equity in global economic development and aid in defeating conventional capitalism are not easy to realize. Again, this is particularly so in developed economies, as strategies to disrupt national economic agendas. In these economies, maintaining business-as-usual is more beneficial as it helps maintain monopolistic positions and is preferable over the trading of their economic privilege to alternative models that would position them at an 'equal playing table' as their political and economic competitors. While calls for a shift from capitalism to socialism are hailed, from an economic standpoint, a radical shift would lead to both dangerous societal aftermaths. It would also equate to a decrease in economic growth, because the competitiveness advocated in the conventional capitalistic model allows for innovation and economic growth. Rafi Khan (2018) posits that the liveability status in many cities has been in direct correlation to economic performance due to capitalism models. This, however, has led to job creation, infrastructural development and advancement in technologies, and is expected to help in the actualization of automation of cities (Robinson, 2018). While building in popularity, socialism, on the other hand, is seen to have a magnetic pull towards low standards of living and reduced innovation (DiLorenzo, 2016; Li, 2017).

Against this background, the question remains as to how to create a more equitable, economic model that would allow continuity of the positive economic activities, while at the same time, addressing the looming societal challenges that have already been brought about by capitalism. With the latter, in our competitive arena, the world is assured of perpetual economic growth, which most developing and less developing economies dearly require if they are to escape the tragedy of climate change especially in regard to infrastructure investment. Through it, the advancement in some areas like technology that have been seen to move to smart technologies is expected to continue growing (Allam, 2020a, 2020b, 2020c, 2020d). On this, though the primary goal may be the accumulation of capital and the maximization of profit by the large Information and communications technology (ICT) corporations, the implementation and installation of these technologies will undoubtedly bring positives in cities through changes in their societal strata, environments and economies (Bahrini & Qaffas, 2019). While socialist principles are commendable and have merits, and their implementation may positively contribute to environmental sustainability, the sluggish growth in economy associated with socialist principles will be detrimental to urban areas facing unprecedented population growth and increased rates of urbanization. For this reason, as is the position of this chapter, a compromise between the two different and established economic models needs to be established and implemented to enable the global economy to continue prospering while addressing the global social, political and environmental challenges.

Re-engineering Marx in the Twenty-First-Century Urban Context

Marx's ideas (Marx & Engels, 2018) principally oriented around the redistribution of wealth with the aim to achieve equality, and fair pricing of labour. In analysing this literature (Hollander, 2008a, 2008b; Screpanti, 2017), and from his own works (Marx, 1973; Marx & Engels, 1885, 2018), it is clear that his propositions focused upon the plight of workers, who according to him have their labour devalued with very low wages. That is, owners of capital and the means of production unfairly compensated labour in their wages. With such de-commensurated efforts of performance and growth, Marx's ideas regarding distribution of wealth are based upon a communist utopia, where Marx, and in extension the Marxists, believed that this economic system has the potential to, first, help abolish the capitalists and secondly, provide a formula for a fair distribution of the wealth (Allam, 2019). This utopic world was characterized as possessing a working environment that was fair and where labour was justly rewarded. It also anticipated a situation where everyone, regardless of their economic status enjoyed equitable housing conditions, whether in cities or in rural areas, that are deemed humane and improved, and therefore they advocated for easy access to the benefits of a modern culture, amongst others (Holt, 2014). According to the Marxists, such opportunities would allow governments to increase their revenue as everyone will be able to contribute taxes according to their abilities. As such, to a certain extent, Marxism may be argued to augur well with modern taxation urban governance models advocating for increased investments and for efficient service deliveries in urban areas.

While some of Marx's ideas were commendable, some were and still have been mis-represented, especially by the political classes and capitalists. The latter have even been responsible for fuelling the communist movement like through the Russian Revolution (1905, 1917) (Hildermeier, 2001), the Chinese public library and Communist movement (1920s and 1930s) (Helling, 2012), the Cuban Revolution (1959) spearheaded by Castro (Moreno Fraginals & Moreno Fraginals, 2001), and in other examples across the world. According to Reynolds (2016), political leaders are not be comfortable with Marxist notions of taxing individuals based upon their economic abilities, and the consequential distribution of this revenue to the needy ones based upon their needs. The problem with this redistribution system is that those being taxed would find creative ways of under-reporting their earnings. On the other hand, recipients of distributed wealth would find ways to exaggerate their needs so they could receive more than what they require. In such cases, if the Marxist's ideas are to hold, it would mean a reduction in economic growth as tax collections would reduce and the number of needy recipients would continue to increase. With insufficient tax revenues, government service deliveries are bound to derail, be overwhelmed, and social problems would continue to increase.

The above negatives, associated with communism, have been recorded to prompt the failure of economies through social uprisings that had subscribed to this idea. While unsurprisingly, these social uprising protagonists cite the extant capitalist movement as a precedent in which countries economically and socially thrive. But, just like Marx had predicted, these patterns have led to a widening gap in social, political and economic inequalities (Juan, 2017; Piketty, 2018). In our modern era, this economic system has also led to unprecedented challenges from an environmental standpoint where a majority of developing and vulnerable economies are facing immense hardships prompted by climate change. This is one of the aftermath consequences of adopting a capitalist economic system, wherein most developed economies have shrewdly exploited its positives.

As supported in the literature (Fankhauser & Stern, 2016; Ward & Mahowald, 2014), most developed economies have been engaging in economic practices that have far too long compromised the integrity of the environment. These activities have greatly impacted upon the economies of the Global South; which is an emerging term used by the World Bank to refer to low- and middle-income countries located in Asia, Africa, Latin America and the Caribbean which contrast to the high-income nations of the Global North. The impacts on the environment by climate change are to the Global South is due to their inadequate infrastructural development that has no capacity to warrant climate change adaptability and resilience. Unfortunately, such economic practices, by developed economies, have also been copied by some of the developing economies outside of the league of former colonial empires. These countries include Brazil, China and India to mention a few, and these countries are even surpassing their developed counterparts in emissions and their compromise of their environment, especially in the case of Brazil (Prevedello, Winck, Weber, Nichols, & Sinervo, 2019; Sonter et al., 2017). In the aforementioned examples, it has been noted that inequalities are taking root, and if such trends continue the concept of social-economic justice will continue to remain a fallacy. It is for this reason that calls towards Marxist ideas have resurfaced and are gaining ground internationally. However, if the commitment regarding the *Paris Agreement* by member economies is adhered to, it will take a considerable amount of time and effort before a shift from conventional capitalism occurs.

As demonstrated in the next section, the slow pace of adopting and integrating Marxism is not a negative rendition. This is because a radical implementation of Marxism can equally lead to dangerous economic outcomes like financial crises, economic stagnation, increased unemployment and increased threats of climate change.

The State vs Disruption: A Case for Socio-Economic Stability

From an economic philosophical perspective, development and economic growth is well achieved when there is unwavering stability, which in part is brought about by pursuing and maintaining a particular economic system. For instance, in this century, the most prominent economic system is that of capitalism, and despite the many glaring shortcomings, its underlying notion is that economies can experience some form of growth. As much as this philosophical perspective remains valid, there are other genuine concerns that cannot wait as they have qualified potential to disrupt the economic growth and development being sought. Some of these concerns are tied to the social and environmental dimensions of the very economies where the concept of capitalism has been accused of undermining. As noted in the sections above, capitalism has been pointed to as one of the main causes of the numerous challenges that are confronting different economies globally. Its emphasis upon growth has been seen to alienate a large group of the people, whom Marx understated in his analysis of capitalist economies. This alienation has exposed them to poverty, unemployment, poor housing, income inequalities, and even worse, their vulnerability to human-induced events of climate change. While these capitalist economies reel in these challenges, those in positions, be it political or in private sectors, have managed to increase their personal wealth, secured lavish and livable environments in country and out country, and enjoy a lifestyle allowing them to financially escape some of the challenges posed by climate change in their home country. Therefore, as much as economic stability is required, there is a need for the urgent revision of existing economic models to facilitate real growth characterized by social justice and economic sustainability.

In terms of social justice and economic sustainability, Roland (2004) calls for caution when contemplating alternative models, since he believes that any drastic revision in existing practices would prompt an abrupt paradigm shift in the economic pursuits of countries and would thereupon have the potential to disrupt and affect extant social strata negatively. Patomäki (2017) argues that this shift would be detrimental to topics like innovation and investment that have been instrumental in pushing economic growth. In particular, it has been observed that most developing and least developed economies have for a long time relied upon Foreign Direct Investment (FDI) to support and grow their economies. But a shift in their economic system would introduce a new risk factor that most investors are concerned about that would be higher thus prompting them to reconsider their investments. As Săvoiu and Ţaicu (2014) highlight, only economies with the potential to afford competitive business environments will manage to attract FDIs and other such investments. A reduction in investments in different sectors will also translate to reduced country government earnings and ultimately, and thereupon a significant reduction in equitable share that Marxists' economic models advocates for.

Ali (2016) further argues that any impact or disruption in economic structures in developing economies reverberates in vulnerable economies as most of these economies depend upon the consumerist trade structures that developed economies exploit. To put this in perspective, the collapse of the former Soviet Union provided an illustration of the importance of a sound and competitive economic system. After the collapse, it has been seen that most of de-Union economies quickly sought FDIs and other investment options to jump-start their own economies (Lane, 2007), and in the process they were able to secure technology and knowledge transfers, and secure larger market bases that allowed them to increase their production as well as create employment opportunities (Cooper, 2013). A drastic disruption in current economic structures will lead to a reverse of these gains, and most economies would mirror the case of the former Soviet economies before the collapse. The worst-case scenario in disrupting the economic systems is that some of the gains that have been achieved in the pursuit of environmental sustainability may end up being trampled upon because most developed economies would equate a change in economic system to an attack upon their economic prosperity. This would lead to jeopardizing their commitments to emission reductions by 2030, as made by most countries, including the financial commitment by developed economies to assist vulnerable economies in climate change mitigation programmes.

With the fear of the above case scenarios materializing, it is then prudent to ensure that any shift in economic systems is undertaken systematically, thus allowing for socio-economic stability. Indeed, the proposition in the *Paris Agreement*, and that of the *SDGs*, provides a substantial timeframe by which particular deliverables, such as emission reductions, can be achieved without drastically changing their country's economic system. On this, Polzer (2019) proposes a gradual mix of different economic systems, merging capitalism and socialism, such that global economies can tap on the strength of both these economic theories, while at the same time avoiding the negative impacts of each. This assertion is augmented by Cole and Ferrarese (2018) who explain that capitalism, as Marx critiques, is more than just about the economy, but it is a form of lifestyle that should be checked to ensure such negatives, like blatant inequalities that emanate from it, are curtailed. These authors assert that entwining it with other economic systems would bring the best out in an economy, and promote an inclusive environment, where the efforts of all are somehow recognized, respected and supported.

Structures of Scale: Transitory Economic Policies

The transition towards a more sustainable and equitable model, with an aim to build a more resilient society, needs to be adaptive to current economic processes so as not to impact negatively upon the economies of vulnerable societies. As showcased in the discussion above, any proposed changes should be systematic and gradual. Thus, they should support a continuous growth trajectory as well as reducing the negatives associated with current systems.

On this front, there are a number of proactive economic strategies that could be adopted to facilitate a smooth transition. The first one, widely discussed by Boussaa (2017), is that urban regeneration has numerous benefits in promoting economic growth, spearheading conservation, and encouraging job creation and new forms of business. Plaza and Haarich (2013) support that through urban regeneration, some cities including Bilbao, Marseille and Rotterdam have managed to forge new economic frontiers that have enabled locals to benefit through job creation and increased business opportunities and also attracted substantial FDIs. Socially, such practices have helped to narrow the inequality gap as locals are financially strengthened.

The second approach is helping local communities and economies tap into their cultural heritage, which according to Vegheş (2018) and Siew and Allam (2017) possesses infinite economic potentials. In the past, due to capitalistic influences, most cultural heritages, especially emanating from poor economies, were seen as backward while those from the former empires and developed economies were celebrated as is evident in the UNESCO World Heritage List inscriptions (Pwiti & Ndoro, 1999; Steiner & Frey, 2011). By exploiting this niche, vulnerable economies can attract economic activities such as tourism that could augment their revenues and reduce their over-reliance on developed economies (Chong & Balasingam, 2018; EY, 2015; Farid, 2015; Jung, Chung, & Leue, 2015; Richards, 2018).

The third strategy places an emphasis upon financing climate change mitigation strategies, especially from green funds (United Nations, 2015a), and the application of financing tools provided in the *Paris Agreement* amongst others (Pan, Elzen, Höhne, Teng, & Wang, 2017; United Nations, 2015b; Ward & Mahowald, 2014). While on this, as has been shown in the literature, that debt financing for the same should be discouraged because it has plunged a sizeable number of economies into financial crisis with some losing, or on the brink of losing, vital public assets.

When adopted and given maximum attention, such strategies have the potential to facilitate the balancing of wealth as was anticipated in Marx's principles. In particular, the ability of such to address societal concerns and to facilitate an increased liveability in urban areas makes the proposed model, as developed in Chapter 4, even more viable. But, as has been discussed above, such need to be supported by developing economies that traditionally defer to capitalists ideologies. The support sought here tasks developed economies to fairly accommodate developing and low-income economies in some of the key decision-making bodies, like UNESCO and Green Funds, amongst others. In addition, to render the paradigm shift more equitable, and for its actualization, the strategies discussed above need to be put in action in both short and medium terms in policy planning so that the challenges faced by vulnerable economies can be addressed now. Also, they should also seek to accommodate long-term planning. Unfortunately, not many models have been seen to prepare us for transitory shifts, thus making the model proposed in this book more valid and of paramount importance.

Conclusion

This concluding chapter explains how it is important to support transitory policy planning at both short and medium terms so that existing economic models are not disrupted thereby not compromising the existing liveability levels of communities. This is of a primordial importance because should global economies be disrupted, this could reverberate on global trade market places and ultimately negatively impact upon vulnerable economies because these economies rely principally upon foreign imports for the survival of their economies. As such, this chapter leads to the conclusion that in our era of capitalism, we need transitory planning measures that can sustain this model while achieving the important dimensions of socialism and environmentalism.

References

Ali, J. A. (2016). Global capitalism and developing countries. *Journal of Competitiveness Studies, 24*(3), 102–106.

Allam, M. Z. (2019). *Urban resilience and economic equity in an era of global climate CRISIS*. Sydney: University of Sydney.

Allam, Z. (2020a). Data as the new driving gears of urbanization. In Z. Allam (Ed.), *Cities and the digital revolution: Aligning technology and humanity* (pp. 1–29). Cham: Springer.

Allam, Z. (2020b). Digital urban networks and social media. In Z. Allam (Ed.), *Cities and the digital revolution: Aligning technology and humanity* (pp. 61–83). Cham: Springer.

Allam, Z. (2020c). On culture, technology and global cities. In Z. Allam (Ed.), *Cities and the digital revolution: Aligning technology and humanity* (pp. 107–124). Cham: Springer.

Allam, Z. (2020d). Privatization and privacy in the digital city. In Z. Allam (Ed.), *Cities and the digital revolution: Aligning technology and humanity* (pp. 85–106). Cham: Springer.

Allam, Z., & Jones, D. (2019). Climate change and economic resilience through urban and cultural heritage: The case of emerging small island developing states economies. *Economies, 7*(2), 62.

Bahrini, R., & Qaffas, A. A. (2019). Technology on economic growth: Evidence from developing countries. *Economies, 7*(21), 1–13.

Boussaa, D. (2017). Urban regeneration and the search for identity in historic cities. *Sustainability, 10*, 48.

Chong, K. Y., & Balasingam, A. S. (2018). Tourism sustainability: Economic benefits and strategies for preservation and conservation of heritage sites in Southern Asia. *Tourism Review, 74*(2), 268–279.

Cole, A., & Ferrarese, E. (2018). How capitalism forms our lives. *Journal for Cultural Research, 22*(2), 105–112.

Cooper, J. (2013). The Russian economy twenty years after the end of the socialist economic system. *Journal of Eurasian Studies, 4*(1), 55–64.

DiLorenzo, T. (2016). *The problem with socialism.* Princeton, NJ: Regnery Publishing.

EY. (2015). *Cultural times: The first global map of culture and creative industries.* Paris, France: EYGM Limited.

Fankhauser, S., & Stern, N. (2016). *Climate change, development, poverty and economics.* Retrieved from http://pubdocs.worldbank.org/en/728181464700790149/Nick-Stern-PAPER.pdf.

Farid, S. M. (2015). Tourism management in world heritage sites and its impact on economic development in Mali and Ethiopia. *Social and Behavioral Sciences, 211,* 595–604.

Helling, J. (2012). 12—China. In J. Helling (Ed.), *Public libraries and their national policies* (pp. 139–147). Cambridge: Chandos Publishing.

Hildermeier, M. (2001). The Russian Revolution. In N. J. Smelser & P. B. Baltes (Eds.), *International encyclopedia of the social & behavioral sciences* (pp. 13433–13437). Oxford: Pergamon.

Hollander, S. (2008a). *The economic of Karl Marx: Analysis and application.* Cambridge, UK: Cambridge University Press.

Hollander, S. (2008b). *The economics of Karl Marx.* Edinburgh Building, Cambridge, UK: Cambridge University Press.

Holt, J. P. (2014). *The social thought of Karl Marx.* London: Sage.

Juan, D. M. M. S. (2017). Why Marx was right: Third world edition. *Journal of Developing Societies, 33*(1), 1–26.

Jung, T., Chung, N., & Leue, M. C. (2015). The determinants of recommendations to use augmented reality technologies: The case of a Korea theme park. *Tourism Management, 49,* 75–86.

Kling, G., Lo, Y., Murinde, V., & Volz, U. (2018). Climate vulnerability and the cost of debt. *SSRN Electronic Journal, 3198093,* 1–30.

Lane, D. (2007). Post-communist states and the European Union. *Journal of Communist Studies and Transition Politics, 23*(4), 461–477.

Li, K.-W. (2017). Chapter 5—Capitalism and socialism: Sustainability versus popularity. In K.-W. Li (Ed.), *Redefining Capitalism in global economic development* (pp. 59–73). Cambridge: Academic Press.

Marx, K. (1973). *Grundrisse.* Harmondsworth: Penguin.

Marx, K., & Engels, F. (1885). *Das Kapital.* Hamburg: Verlag von Otto Meissner.

Marx, K., & Engels, F. (2018). *The communist Manifesto*. Minneapolis, MN: Lerner Publishing Group Inc.

Moreno Fraginals, M. R., & Moreno Fraginals, T. P. (2001). The Cuban revolution. In N. J. Smelser & P. B. Baltes (Eds.), *International encyclopedia of the social & behavioral sciences* (pp. 3027–3032). Oxford: Pergamon.

Pan, X., Elzen, M. D., Höhne, N., Teng, F., & Wang, L. (2017). Exploring fair and ambitious mitigation contributions under the Paris Agreement goals. *Environmental Science & Policy, 74,* 49–56.

Patomäki, H. (2017). Capitalism: Competition, conflict, crisis. *Journal of Critical Realism, 16*(5), 537–543.

Piketty, T. (2018). *Top incomes in France in the twentieth century: Inequality and redistribution, 1901–1998*. Cambridge: Harvard University Press.

Plaza, B., & Haarich, S. N. (2013). The Guggenheim Museum Bilbao: Between regional embeddedness and global networking. *European Planning Studies, 23*(8), 1456–1475.

Polzer, K. (2019). *Mixing capitalism and socialism: Policies to moderate systemic wealth concentration in the United States.*

Prevedello, J. A., Winck, G. R., Weber, M. M., Nichols, E., & Sinervo, B. (2019). Impacts of forestation and deforestation on local temperature across the globe. *PLoS One, 14*(3), e0213368.

Pwiti, G., & Ndoro, W. (1999). The legacy of colonialism: Perceptions of the cultural heritage in Southern Africa, with special reference to Zimbabwe. *The African Archaeological Review, 16*(3), 143–153.

Rafi Khan, S. (2018). Reinventing capitalism to address automation: Sharing work to secure employment and income. *Competition & Change, 22*(4), 343–362.

Reynolds, A. (2016, February 15). *What Marx got right about redistribution—That John Stuart Mill got wrong*. Retrieved from https://fee.org/articles/marx-was-right-about-redistribution/.

Richards, G. (2018). Cultural Tourism: A review of recent research and trends. *Journal of Hospitality and Tourism Management, 36,* 12–21.

Robinson, W. I. (2018). The next economic crisis: Digital capitalism and global police state. *Race & Class, 60*(1), 77–92.

Roland, G. (2004). Understanding institutional change: Fast-moving and slow-moving institutions. *Studies in Comparative International Development, 38*(4), 109–131.

Săvoiu, G., & Ţaicu, M. (2014). Foreign direct investment models, based on country risk for some post-socialist central and eastern european economies. *Procedia Economics and Finance, 10,* 249–260.

Screpanti, E. (2017). Karl Marx on wage labor: From natural abstraction to formal subsumption. *Rethinking Marxism, 29*(4), 511–537.

Siew, G., & Allam, Z. (2017). *Culture as a driver for sustainable urban development.* Paper presented at the UIA 2017 Seoul World Architects Congress, Seoul. http://www.uia2017seoul.org/P/papers/Abstract/Paper/Oral/PS2-52/O-0091.pdf.

Sonter, L. J., Herrera, D., Barrett, D. J., Galford, G. L., Moran, C. J., & Soares-Filho, B. S. (2017). Mining drives extensive deforestation in the Brazilian Amazon. *Nation Communication, 8*(1), 1013.

Steiner, L., & Frey, B. S. (2011). *Imbalance of world heritage list: Did the UNESCO strategy work?* Zurich. Retrieved from http://www.econ.uzh.ch/static/wp/econwp014.pdf.

Tol, R. S. J. (2018). The economic impacts of climate change. *Review of Environmental Economics and Policy, 12*(1), 4–25.

United Nations. (2015a). *Addis Ababa Action Agenda of the Third International Conference on Financing for Development (Addis Ababa Action Agenda).* Paper presented at the Third International Conference on Financing for Development Addis Ababa, Ethiopia.

United Nations. (2015b). *Resolution adopted by the General Assembly on 25th September 2015.* Retrieved from https://www.un.org/en/development/desa/population/migration/generalassembly/docs/globalcompact/A_RES_70_1_E.pdf.

Vegheş, C. (2018). Cultural heritage, sustainable development and inclusive growth: Global lessons for the local communities under a marketing approach. *European Journal of Sustainable Development, 7*(4), 349–360.

Ward, D. S., & Mahowald, N. M. (2014). Contributions of developed and developing countries to global climate forcing and surface temperature change. *Environmental Research Letters, 9*(7), 074008.

Wright, H., Reeves, J., & Huq, S. (2016). *Impact of climate change on least developed countries: Are the SDGs possible?* https://doi.org/10.13140/rg.2.1.2817.0649.

Index

Z. Allam et al., *Cities and Climate Change*, Palgrave Studies in Climate Resilient Societies, https://doi.org/10.1007/978-3-030-40727-8

The manufacturer's authorised representative in the EU is Springer
Nature Customer Service Centre GmbH, Europaplatz 3, 69115 Heidelberg,
Germany. If you have any concerns regarding our products, please
contact ProductSafety@springernature.com

Printed and bound by CPI Group (UK) Ltd, Croydon, CR0 4YY
25/09/2026
02222894-0006